STARTING A CATFISH FARM

An In-Depth Guide to Starting and Managing Your Catfish Farm, Covering Pond Design, Feeding Techniques, Harvesting Methods, Disease Prevention and Sustainable Practices.

FRANK S. COUCH

All rights reserved. No part of this publication may be reproduced, stored in a retrieval system, or transmitted in any form or by any means, electronic, mechanical, photocopying, recording, or otherwise, without the prior written permission of the publisher, except for brief quotations in critical reviews or articles.

Copyright © 2024 by Frank S. Couch.

CONTENTS

Introduction — 5
  What is Catfish Farming? — 5
  Why Choose Catfish Farming? — 8
  Who Should Start a Catfish Farm? — 12

CHAPTER 1 — 15

Understanding Catfish — 15
  1.1 Types of Catfish for Farming — 15
  1.2 Life Cycle of Catfish — 22
  1.3 Choosing the Right Breed for Your Farm — 28

CHAPTER 2 — 32

Planning Your Catfish Farm — 32
  2.1 Setting Goals for Your Catfish Farm — 32
  2.2 Selecting a Suitable Location for Your Catfish Farm — 37
  2.3 Legal and Regulatory Requirements for Catfish Farming — 42

CHAPTER 3 — 48

Building the Infrastructure — 48
  3.1 Types of Farming Systems in Catfish Farming — 48
  3.2 Pond Construction and Design for Catfish Farming — 54
  3.3 Setting Up Tanks and Equipment for Catfish Farming — 59

CHAPTER 4 — 66

Procuring and Stocking Fingerlings — 66
  4.1 Where to Buy Healthy Fingerlings for Your Catfish Farm — 66
  4.2 Transporting Fingerlings Safely — 70
  4.3 Stocking Your Pond or Tanks — 75

CHAPTER 5 — 79

Feeding and Nutrition — 79
  5.1 Understanding Catfish Dietary Needs — 79
  5.2 Feeding Schedules and Techniques — 84
  5.3 Managing Feed Costs and Efficiency — 88

CHAPTER 6 — 93

Managing Water Quality — 93
- 6.1 Importance of Water Quality in Catfish Farming — 93
- 6.2 Regular Water Testing — 96
- 6.3 Common Water Quality Problems and Solutions — 100

CHAPTER 7 — 104

Health Management and Disease Prevention — 104
- 7.1 Identifying Common Catfish Diseases — 104
- 7.2 Disease Prevention Strategies — 107
- 7.3 Treatment and Recovery — 111

CHAPTER 8 — 115

Harvesting and Marketing — 115
- 8.1 When to Harvest Your Catfish — 115
- 8.2 Harvesting Techniques — 118
- 8.3 Selling Your Catfish — 122

CHAPTER 9 — 126

Financial Management and Sustainability — 126
- 9.1 Budgeting for Your Catfish Farm — 126
- 9.2 Managing Your Profit Margins — 129
- 9.3 Sustainable Catfish Farming Practices — 133

Conclusion — 138
- Challenges to Expect in Catfish Farming and How to Overcome Them — 138
- Encouragement for Your Catfish Farming Journey — 141

INTRODUCTION

What is Catfish Farming?

Catfish farming, also known as catfish aquaculture, is the practice of raising catfish in controlled environments such as ponds, tanks, or recirculating aquaculture systems (RAS) for commercial, personal, or subsistence purposes. This method of farming involves breeding, feeding, managing water quality, and harvesting catfish to produce a reliable source of protein for human consumption.

Catfish are a highly popular and valuable species for aquaculture due to several key characteristics. First, they are hardy fish that can adapt to a wide range of environments, making them relatively easy to farm. They grow quickly, have a high feed-to-growth conversion ratio, and are resistant to many diseases. Additionally, catfish are a staple food source in many regions around the world, making their farming a lucrative business.

How Does Catfish Farming Work?

Catfish farming typically starts with the selection of suitable fingerlings (young catfish) or brood stock (adult catfish used for breeding). These are placed in a controlled environment where they are provided with the right conditions for growth. Farmers closely monitor the water quality, feed the catfish a nutritionally balanced diet, and ensure the fish are healthy and growing efficiently.

After a period of growth, which can last several months depending on the size of the farm and the market demand, the catfish are harvested. At this point, they can either be sold live, processed into fillets, or turned into other fish products for sale at markets, restaurants, or grocery stores.

Components of Catfish Farming

- Ponds or Tanks: Catfish are typically raised in earthen ponds, but more advanced systems like tanks and recirculating aquaculture systems (RAS) are becoming more common. Ponds are generally favoured for large-scale operations, while tanks or RAS systems are ideal for smaller farms or urban areas with limited space.

- Fingerlings: The process begins with stocking ponds or tanks with fingerlings. Fingerlings are young catfish, usually a few inches long, that will grow into harvestable fish. Sourcing healthy fingerlings from reputable hatcheries is critical for the success of the farm.

- Feeding: Catfish have specific dietary needs to promote optimal growth. Farmers use high-protein commercial feed designed for catfish, which floats on the water surface to ensure the fish eat efficiently. Feeding schedules are carefully managed to maximize growth and minimize waste.

- Water Management: Proper water quality is crucial to catfish farming. Farmers need to monitor and manage key water parameters such as oxygen levels, pH, temperature, and ammonia content. Without proper water management, fish health can quickly deteriorate, leading to disease outbreaks or poor growth.

- Harvesting: Once the catfish reach market size, which usually takes 6 to 12 months depending on the species, they are harvested. The method of harvesting can vary depending on the system, but it typically involves draining ponds or using nets to gather the fish.

- Marketing and Sales: After harvesting, the catfish are either sold live or processed into fillets and other products. Farmers can market their catfish locally to restaurants, fish markets, or even grocery stores. Some farms also process and freeze the fish for wider distribution.

Why is Catfish Farming Important?

- Economic Benefits: Catfish farming provides an affordable source of income for many people, particularly in areas where agriculture is the primary industry. In some countries, catfish farming has grown into a multi-million-dollar industry, supporting thousands of jobs and boosting local economies.

- Sustainability: Catfish farming is considered one of the most sustainable forms of aquaculture because catfish are highly efficient at converting feed into body mass. This means they require less feed to grow, reducing the environmental footprint of fish farming. Furthermore, with proper management, catfish farming can be done with minimal impact on natural water resources.

- Food Security: As the global population grows, there is increasing demand for reliable, affordable protein sources. Catfish farming helps meet this demand by providing a steady supply of nutritious, protein-rich food. Because catfish are relatively easy to farm and can be raised in controlled environments, they offer a dependable food source in regions with limited access to wild fish or other protein sources.

- Environmental Benefits: In comparison to wild fishing, which can deplete natural fish populations, catfish farming allows for controlled fish production without putting pressure on wild ecosystems. Additionally, modern farming techniques, such as recirculating aquaculture systems, use minimal water and can be managed with little environmental disruption.

Catfish farming is an accessible and profitable form of aquaculture, ideal for individuals and communities looking to produce a sustainable and reliable source of food. Whether done on a small scale in tanks or a large commercial scale in ponds, the fundamentals of raising catfish remain the same: providing proper nutrition, managing water quality, and ensuring a healthy growing environment. For anyone interested in entering the world of aquaculture, catfish farming offers a promising opportunity to contribute to food security while enjoying the economic benefits of this thriving industry.

Why Choose Catfish Farming?

Catfish farming is one of the most appealing options for anyone interested in aquaculture, whether you're looking to start a profitable business or simply provide a reliable source of food for your community. With its numerous advantages, it stands out as a versatile and sustainable farming practice. Here are several key reasons why choosing catfish farming can be a smart and rewarding choice.

1. High Demand and Market Potential

One of the most compelling reasons to choose catfish farming is the consistent demand for catfish as a food source. Across the world, particularly in regions like the United States, Africa, and Southeast Asia, catfish is a popular and affordable protein. It is widely used in various cuisines and is often the preferred fish for grilling, frying, and stewing.

With an increasing global population and growing demand for affordable and sustainable seafood, the market potential for catfish farming continues to expand. Many restaurants, grocery stores, and fish markets seek reliable suppliers of fresh or processed catfish, making it an attractive option for aspiring fish farmers.

Additionally, catfish farming is adaptable to both local and international markets, offering flexibility for those who want to scale their operations over time. In some areas, there are opportunities to supply not only fresh fish but also value-added products such as catfish fillets, smoked catfish, and even fish by-products used in animal feeds and fertilizers.

2. Low Start up and Operational Costs

Compared to other forms of animal farming or aquaculture, catfish farming has relatively low start up and operational costs. You don't need extensive land or expensive equipment to get started. With just a small pond or a few tanks, beginners can raise catfish and gradually expand their farm as they gain experience.

Catfish thrive in a wide range of environments, which makes setting up a farm more affordable and flexible. Depending on your location, you can choose between traditional pond systems or opt for tank-based systems if space is limited. The essential infrastructure needed—such as aerators for oxygen supply, nets for harvesting, and basic feeding equipment—doesn't require a large financial investment.

In terms of operational costs, feeding catfish is relatively cost-effective. Catfish have a high feed-to-growth conversion rate, meaning they require less feed compared to other farmed animals to achieve optimal growth. This efficiency helps keep costs low while maximizing production.

3. Fast Growth and High Yield

Catfish farming offers an advantage that appeals to both small-scale and commercial farmers: catfish grow quickly and can be harvested in a relatively short period. Under the right conditions, catfish can grow to market size in about 6 to 12 months, depending on the species and feeding regime.

This fast growth cycle means farmers can harvest more frequently, providing a continuous income stream. Additionally, catfish farming can be done in relatively high densities, meaning you can raise a large number of fish in a small space without compromising their health or growth. This makes it an efficient way to use available land or water resources.

For farmers who want to scale up production, catfish farming offers a clear path to expansion. With proper planning, you can increase the size of your farm and your yields over time, making it a sustainable and scalable business.

4. Hardy Species with High Disease Resistance

Catfish are known for being hardy, adaptable fish that can survive in diverse environmental conditions. This resilience makes catfish farming more forgiving, especially for beginners who may not have extensive experience with fish farming.

Unlike some other species, catfish are relatively resistant to many common diseases and parasites. With proper water quality management and a good feeding regimen, the risk of disease outbreaks can be kept to a minimum. Even when problems do arise, catfish typically respond well to treatment, allowing farmers to maintain healthy stock and avoid large-scale losses.

This hardiness also means that catfish farming can thrive in various climates and regions, from tropical areas to temperate zones. Whether you're farming in outdoor ponds or controlled indoor systems, catfish can adapt well and continue growing efficiently.

5. Efficient Feed Conversion

Another major reason to choose catfish farming is the species' excellent feed conversion ratio (FCR). In aquaculture, FCR is the measure of how much feed is required to produce a certain amount of fish biomass. Catfish are known for their efficient use of feed, meaning they convert a greater percentage of the feed they consume into body mass.

For example, while some fish species require a high amount of protein-rich feed to grow, catfish can thrive on relatively inexpensive feed formulations. They can digest a wide range of plant-based and animal-based feed, which helps to keep feeding costs low. This ability to efficiently convert feed into growth allows catfish farmers to optimize their operations and reduce waste.

6. Simple Farming Techniques

Catfish farming doesn't require advanced technical knowledge or specialized equipment, making it an accessible venture for beginners. Whether you choose to start with a small backyard pond or a more structured tank system, the principles behind catfish farming are easy to learn and follow.

Farmers primarily need to focus on a few key areas:

- Water quality management (monitoring oxygen levels, pH, temperature, and ammonia levels)
- Feeding (providing nutritionally balanced feed on a regular schedule)
- Health monitoring (observing the fish for signs of disease or stress)

By mastering these basic aspects of catfish farming, beginners can achieve success without needing years of training or expensive consulting services.

7. Sustainable and Environmentally Friendly

In today's world, sustainability is an important consideration for any farming practice. Catfish farming is considered one of the most environmentally friendly forms of aquaculture. Here's why:

- Water use efficiency: Modern catfish farms, especially those using recirculating aquaculture systems (RAS), use water very efficiently. By recycling and filtering water, farmers can minimize water consumption while maintaining optimal conditions for their fish.

- Low environmental impact: Unlike traditional wild fishing, which can lead to overfishing and depletion of natural fish populations, catfish farming is a controlled method of fish production. Farmers can raise large numbers of fish without affecting wild ecosystems.

- Reduced waste: Because catfish have such efficient feed conversion ratios, there is less waste in the form of uneaten feed or excreted material, reducing the environmental impact of fish farming.

Additionally, catfish farms can be integrated with other forms of agriculture. For example, pond water rich in nutrients from fish waste can be used to irrigate crops, creating a closed-loop system that benefits both the fish and plant growth.

8. Job Creation and Community Development

Catfish farming offers the potential for job creation and community development, especially in rural areas where economic opportunities may be limited. As the farm grows, it creates jobs for farmhands, technicians, and workers involved in processing and marketing the fish.

In some regions, catfish farming has become a key part of local economies, supporting thousands of families and helping to reduce poverty. By investing in a catfish farm, you not only build a profitable business for yourself but also contribute to the economic health of your community.

Choosing catfish farming offers numerous advantages, from its high market demand and low start up costs to the fast growth of the fish and the sustainability of the farming practices. Whether you're looking to start small or scale up to a commercial operation, catfish farming provides a rewarding opportunity to grow a profitable business while contributing to food security and environmental sustainability. For beginners and experienced farmers alike, catfish farming is a practical, profitable, and environmentally responsible choice.

Who Should Start a Catfish Farm?

Starting a catfish farm can be a rewarding and profitable venture, but it's important to know if it's the right fit for you. Whether you're a beginner looking to break into aquaculture or an established farmer seeking to diversify your business, catfish farming has opportunities for a wide range of individuals. Below is an extensive look at who should consider starting a catfish farm.

1. Aspiring Aquaculture Entrepreneurs

If you're someone looking to enter the world of aquaculture for the first time, catfish farming is an excellent starting point. It offers several advantages that make it ideal for beginners, including low start up costs, simple farming techniques, and fast-growing fish.

Catfish farming doesn't require a degree in marine biology or years of aquaculture experience. With some basic training or self-study, anyone with a passion for fish farming and a willingness to learn can succeed in this field. In fact, many successful catfish farmers began as complete novices, using small ponds or tanks to get their feet wet before scaling up operations.

2. Small-Scale Farmers or Hobbyists

If you already have a piece of land, such as a small farm, or even a large backyard, you're in an excellent position to start a small-scale catfish farming operation. Many farmers begin with a few ponds or tanks to supplement their existing income or to produce food for their families and communities.

Hobbyists who enjoy working with animals, especially those who already raise livestock or grow crops, can find catfish farming to be a complementary addition to their agricultural practices. It's a way to diversify your income without needing extensive new infrastructure, and it can also be integrated with other farming activities, like using nutrient-rich water from the ponds to irrigate crops.

3. Commercial Farmers Looking to Diversify

For established farmers who primarily focus on crops or livestock, catfish farming is a great way to diversify your business. The fish farming industry is rapidly growing due to increasing demand for sustainable and locally produced seafood. Adding catfish to your farm can provide a steady income stream, reduce risk, and make your farm more resilient in the face of market or environmental fluctuations.

Additionally, many farmers already have the necessary land or water resources to set up ponds or tanks, making catfish farming an easy addition. With proper planning and resource allocation, you can scale catfish farming alongside your other operations without needing significant new investments.

4. People in Rural or Underserved Communities

If you live in a rural or underserved area, starting a catfish farm can be an opportunity to create both personal and community-wide benefits. Catfish farming can provide a reliable source of income in areas where traditional employment opportunities are limited. Additionally, it can help increase food security by producing a steady supply of affordable, locally sourced protein.

Many rural communities have access to natural bodies of water or inexpensive land, making them well-suited for catfish farming. Local governments and non-governmental organizations (NGOs) may also offer grants or financial assistance to those interested in starting farms, particularly in regions where food production is a priority.5. Environmentally Conscious Entrepreneurs

For those who are passionate about sustainability and environmentally friendly business practices, catfish farming offers a way to produce food with a low environmental impact. Unlike other types of animal farming, which often require large amounts of land, water, and feed, catfish farming can be done efficiently in smaller spaces. Catfish are also efficient at converting feed into body mass, reducing waste and making them a sustainable source of protein.

Additionally, with modern practices like recirculating aquaculture systems (RAS), which recycle and clean water within the farm, it's possible to raise catfish while using minimal natural resources. If you're looking to start a business that prioritizes environmental sustainability, catfish farming can be a great choice.

6. Individuals Seeking a Steady Income or Self-Sufficiency

If you're someone who values self-sufficiency or wants to generate a stable, reliable income, catfish farming can offer both. Once the farm is up and running, catfish provide a consistent supply of fish that can be sold to markets, restaurants, or directly to consumers. The fast growth rate of catfish also means that you don't have to wait long to start seeing returns on your investment.

Catfish farming can be managed as a full-time venture or as a side business to generate additional income. For those who wish to feed their own family, it offers a steady source of fresh, healthy fish without having to rely on expensive grocery store options.

7. People with Access to Land or Water Resources

For individuals who already have access to land or water, such as farmers, landowners, or rural residents, catfish farming can be a perfect way to utilize those resources. You don't need a vast amount of land to start a catfish farm; even a small pond or a series of tanks can produce a substantial number of fish.

Those with access to natural water bodies like ponds, rivers, or reservoirs can easily set up a low-cost catfish farming operation. Additionally, individuals living near these resources can explore partnerships or lease agreements with landowners to create mutually beneficial farming arrangements.

Catfish farming is a versatile and rewarding endeavour suitable for a wide range of individuals. Whether you're an aspiring entrepreneur, an established farmer, or someone interested in sustainable and profitable farming practices, catfish farming offers numerous benefits. Its low start up costs, high demand, and adaptability makes it an attractive option for beginners and experienced farmers alike.

If you have access to land or water, a desire for self-sufficiency, or a passion for sustainable farming, catfish farming may be the perfect venture for you. With dedication and the right knowledge, anyone can succeed in this growing and lucrative industry.

CHAPTER 1

UNDERSTANDING CATFISH

1.1 Types of Catfish for Farming

When it comes to starting a catfish farm, one of the most important decisions you'll make is choosing the right species of catfish. Different types of catfish have varying characteristics, including growth rate, adaptability to different environments, feeding habits, and disease resistance. Selecting the right type for your farm can significantly impact your success. Below is an extensive, reader-friendly explanation of the most common types of catfish used in farming:

1. Channel Catfish

Channel catfish (Ictalurus punctatus) is the most popular species for farming, especially in the United States and other parts of North America. Known for its rapid growth, high meat yield, and adaptability, it is the go-to species for most commercial catfish farms.

Characteristics:

- Growth Rate: Channel catfish grow quickly under the right conditions, reaching market size (usually around 1.5 to 2 pounds) within 18 to 24 months.

- Adaptability: They are highly adaptable to various environments, thriving in both ponds and tank-based farming systems.

- Feeding: They are omnivorous, which means they can eat a wide variety of foods, including formulated pellets, making them easier and more affordable to feed.

- Disease Resistance: Channel catfish are relatively hardy and can withstand common diseases better than some other species, although they are still vulnerable to parasites and bacterial infections.

- Temperature Tolerance: They prefer warm water and grow best in temperatures between 75°F and 85°F (24°C to 29°C), making them suitable for warm climates.

Why Choose Channel Catfish?

- Commercial Viability: Channel catfish are the backbone of the catfish farming industry due to their high market demand and suitability for large-scale farming.

- Ease of Farming: Their ability to adapt to various farming systems and tolerate less-than-ideal water conditions makes them ideal for both beginners and experienced farmers.

2. African Catfish

The African catfish (Clarias gariepinus) is widely farmed across Africa and is gaining popularity in other regions due to its hardy nature, fast growth, and high resistance to poor water quality.

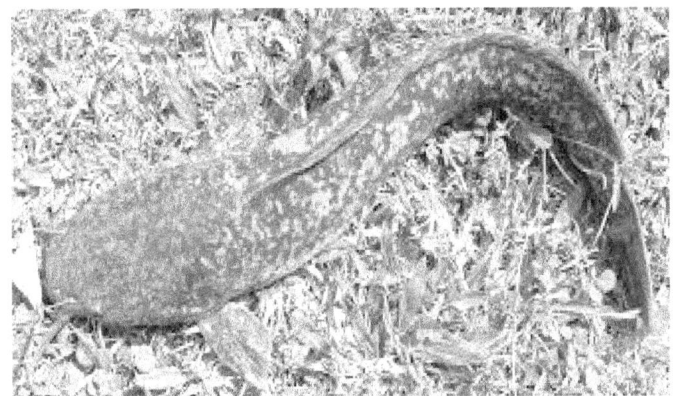

Characteristics:

- Growth Rate: African catfish grow rapidly, often faster than channel catfish, reaching market size within 6 to 12 months in optimal conditions.

- Hardiness: One of the standout features of African catfish is their ability to survive in low-oxygen environments. This makes them suitable for farming in areas where water quality or oxygen levels may be a concern.

- Feeding: They are also omnivorous, eating a variety of foods including fish meal, plant material, and artificial feed.

- Disease Resistance: African catfish are incredibly resilient and can resist many of the diseases and parasites that affect other species.

- Temperature Tolerance: They thrive in warm temperatures, ideally between 77°F and 86°F (25°C to 30°C), making them ideal for tropical or subtropical climates.

Why Choose African Catfish?

- Resilience: If you live in an area with poor water quality or frequent temperature fluctuations, African catfish are a great choice due to their hardiness and ability to survive challenging conditions.

- Rapid Growth: Their quick growth means you can produce market-ready fish faster, which translates to quicker returns on your investment.

- Ideal for Developing Regions: African catfish are a popular choice for farming in developing countries because of their resilience, even in less-than-ideal farming conditions.

3. Blue Catfish

The blue catfish (Ictalurus furcatus) is another popular species, especially in the United States. It is similar to the channel catfish but has some key differences that may make it more appealing in certain farming situations.

Characteristics:

- Growth Rate: Blue catfish grow larger than channel catfish and can reach market size within a similar time frame, but they can grow significantly bigger—up to 100 pounds in the wild.

- Adaptability: While they are adaptable, blue catfish prefer flowing water and may require more complex pond or tank systems to thrive.

- Feeding: Like channel catfish, blue catfish are omnivorous and can be fed formulated diets or natural food sources found in ponds.

- Disease Resistance: Blue catfish are generally more resistant to some of the diseases that affect channel catfish, making them a healthier option for certain farms.

- Temperature Tolerance: They prefer warmer waters but can tolerate cooler conditions better than channel catfish, making them suitable for a broader range of climates.

Why Choose Blue Catfish?

- Larger Size: If you're interested in producing larger catfish, blue catfish offer the potential to grow much bigger than channel catfish.

- Disease Resistance: Their higher resistance to disease can reduce costs associated with medication and improve survival rates in your farm.

- Versatility: Blue catfish can handle a wider range of temperatures, which means they can be farmed in areas where other species may struggle.

4. Hybrid Catfish

Hybrid catfish are a cross between channel catfish and blue catfish. This hybrid species combines the best traits of both parents, making it a highly popular choice among catfish farmers.

Characteristics:

- Growth Rate: Hybrid catfish grow faster than either parent species, which means they reach market size quicker, often within 12 to 18 months.

- Hardiness: Hybrids inherit the disease resistance of blue catfish and the adaptability of channel catfish, making them a strong, resilient choice for farming.

- Feeding: Like both parents, hybrids are omnivorous and can thrive on formulated feeds, making them relatively easy to manage.

- Survival Rates: Hybrid catfish tend to have higher survival rates than either pure channel or blue catfish, especially in farm conditions.

- Temperature Tolerance: Hybrids thrive in warm climates, but their resilience makes them suitable for a variety of environmental conditions.

Why Choose Hybrid Catfish?

- Fast Growth: If you're looking to maximize your profits, hybrid catfish are ideal because they grow faster and reach market size quicker.

- High Survival Rates: The hardiness of hybrid catfish means fewer losses due to disease or environmental stress, leading to a more profitable farm.
- Best of Both Worlds: By combining the best traits of channel and blue catfish, hybrids are a perfect choice for farmers looking for a versatile, fast-growing, and resilient species.

5. Walking Catfish

The walking catfish (Clarias batrachus) is an interesting species primarily found in Asia. They are known for their unique ability to "walk" on land for short distances, thanks to their strong pectoral fins.

Characteristics:

- Growth Rate: Walking catfish grow quickly in warm, tropical environments and can reach market size within 6 to 12 months.
- Adaptability: They can survive in low-oxygen water conditions and are highly adaptable to extreme environments.
- Feeding: Walking catfish are omnivorous and can be fed a wide range of diets, including natural food sources and formulated feeds.
- Hardiness: Their ability to survive in harsh conditions, including the ability to migrate short distances on land, makes them one of the hardiest catfish species.

- Temperature Tolerance: They thrive in tropical climates, with a preferred temperature range of 77°F to 86°F (25°C to 30°C).

Why Choose Walking Catfish?

- Extreme Hardiness: If you live in an area with challenging environmental conditions, walking catfish are an excellent choice due to their ability to thrive in low-oxygen and poor water conditions.

- Niche Market: Walking catfish can cater to specific markets, particularly in Asia, where they are in high demand. They are also considered a delicacy in certain cuisines, making them a good option for farmers targeting niche markets.

Choosing the right type of catfish for farming is a crucial step that depends on your local climate, available resources, and market demands. Channel catfish are ideal for beginners and commercial farmers alike due to their adaptability and wide market acceptance. African catfish are perfect for those in regions with poor water quality, while blue catfish offer larger sizes and better disease resistance. If you're looking for a fast-growing, resilient option, hybrid catfish provide the best traits of both channel and blue catfish. Walking catfish, though niche, offer hardiness and are suited for specific markets. By understanding the different types of catfish and their unique qualities, you can make an informed decision that maximizes your chances of success as a catfish farmer.

1.2 Life Cycle of Catfish

Understanding the life cycle of catfish is crucial for successfully managing a catfish farm. From egg to adult, catfish go through several key stages that require specific care and attention. Knowing the needs and behaviours of catfish at each life stage will help you optimize growth, maintain healthy stock, and increase your farm's productivity. Here's a detailed explanation of the different stages of the catfish life cycle:

1. Spawning (Egg Stage)

The life cycle of a catfish begins when mature adults, often referred to as broodstock, spawn (lay eggs). Catfish typically spawn in warm water when the temperature reaches around 75°F to 85°F (24°C to 29°C), which is why most catfish farms focus on warm climates or maintain heated environments during breeding.

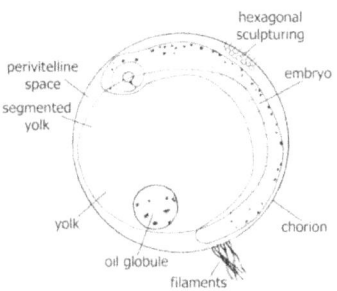

How Spawning Happens:

- Broodstock Selection: Healthy male and female catfish (usually 2 to 4 years old) are selected for breeding. A well-prepared broodstock ensures high-quality eggs and strong offspring.

- Nesting: Catfish prefer to lay their eggs in protected, dark areas, such as submerged cavities or artificial structures like spawning boxes provided in ponds or tanks. In the wild, they may use hollow logs or rocks.

- Egg Deposition: The female lays thousands of eggs in a gelatinous mass, attaching them to the walls of the nesting area. The male fertilizes the eggs shortly after they are laid.

- Incubation Period: The eggs incubate in the nest for about 5 to 10 days, depending on the water temperature. Warmer temperatures speed up the incubation process. During this time, the male catfish guards the nest to protect the eggs from predators.

Considerations for Farmers:

- Water Quality: Maintain clean, well-oxygenated water in the spawning areas to prevent fungal infections and other issues that could damage the eggs.
- Protection: Provide safe and secure nesting areas for the broodstock to encourage successful spawning.

2. Hatching (Larval Stage)

After about a week, the eggs hatch into larvae, also known as fry. This stage is critical because the newly hatched fry are highly vulnerable and require special care to ensure they survive and grow into healthy juveniles.

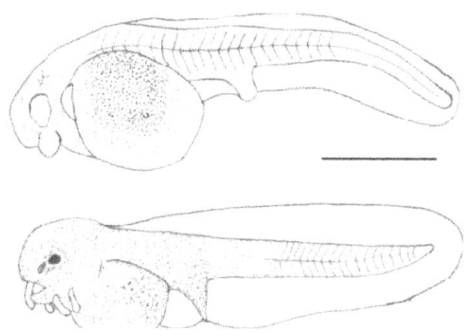

What Happens During Hatching:

- Yolk Sac Absorption: After hatching, the fry are not yet ready to feed on their own. They survive by absorbing the nutrients from their attached yolk sacs for about 3 to 5 days.
- Feeding Begins: Once the yolk sac is fully absorbed, the fry must start feeding. In the wild, they eat small aquatic organisms, while in farms, they are often fed small, nutrient-rich feed designed specifically for fry.

Considerations for Farmers:

- Feeding: Start feeding the fry with specially formulated feeds that are easy for them to digest. Proper nutrition at this stage is essential for fast growth and survival.

- Protection: Fry are extremely delicate, so protect them from predators, strong water currents, and poor water conditions.
- Water Quality: Ensure that the water is clean and well-oxygenated, as poor water quality can quickly lead to high mortality rates at this stage.

3. Fingerling Stage (Juvenile Stage)

Once the fry have grown large enough to be called fingerlings, they begin to develop the traits of adult catfish. This is the stage where they experience rapid growth, making it a crucial period for farm productivity.

What Happens During the Fingerling Stage:

- Rapid Growth: Fingerlings grow quickly under the right conditions. At this stage, they range from 2 to 6 inches in length and begin to resemble miniature versions of adult catfish.
- Feeding: At this point, they can handle larger feed pellets that are high in protein. Frequent, controlled feeding helps them grow fast and reduces competition for food.
- Stocking: This is the time when fingerlings are often moved from hatchery tanks or protected nurseries into larger grow-out ponds or tanks, where they will continue to grow into adults.

Considerations for Farmers:

- Stocking Density: Overcrowding can lead to poor growth, stress, and disease outbreaks. Keep a balanced stocking density in your grow-out ponds to allow the fingerlings to grow comfortably.

- Feeding Program: Implement a consistent feeding schedule to ensure uniform growth. Use high-quality, nutrient-dense feed to support their rapid growth phase.

- Water Management: Keep a close eye on water quality, as the increasing biomass (the number of fish) will affect oxygen levels and waste build up in the pond.

4. Adult Stage (Grow-Out Phase)

In the grow-out phase, catfish reach sexual maturity and continue to grow until they are ready for harvest. Depending on the species and farming conditions, catfish typically reach market size at around 1.5 to 2 pounds within 18 to 24 months for species like channel catfish. However, this time frame may vary depending on the type of catfish and farming practices.

What Happens During the Adult Stage:

- Growth Plateau: After rapid growth in the fingerling stage, the growth rate of catfish slows as they approach maturity. However, regular feeding and proper management can still help them reach their full market size.

- Harvesting: Once they have reached the desired market size, the catfish are ready for harvesting. Most farmers harvest their catfish when they are between 1 to 3 pounds, although some species, like blue catfish, can grow much larger if left in the pond longer.

Considerations for Farmers:

- Feeding Efficiency: As fish grow, their feed requirements change. Adjust your feeding schedule to optimize growth and minimize feed waste.

- Health Monitoring: Adult catfish can be susceptible to diseases, parasites, and environmental stressors. Monitor your stock regularly for signs of illness and maintain proper water quality to prevent health issues.

- Harvesting Techniques: Plan your harvest carefully to minimize stress and injury to the fish. Stress during harvest can affect the quality of the meat and overall yield.

5. Broodstock (Mature Stage)

For farmers interested in breeding their own catfish, a small number of adult catfish can be kept as broodstock. These mature fish are selected based on their size, health, and breeding potential. The broodstock will be used to produce the next generation of fish for the farm.

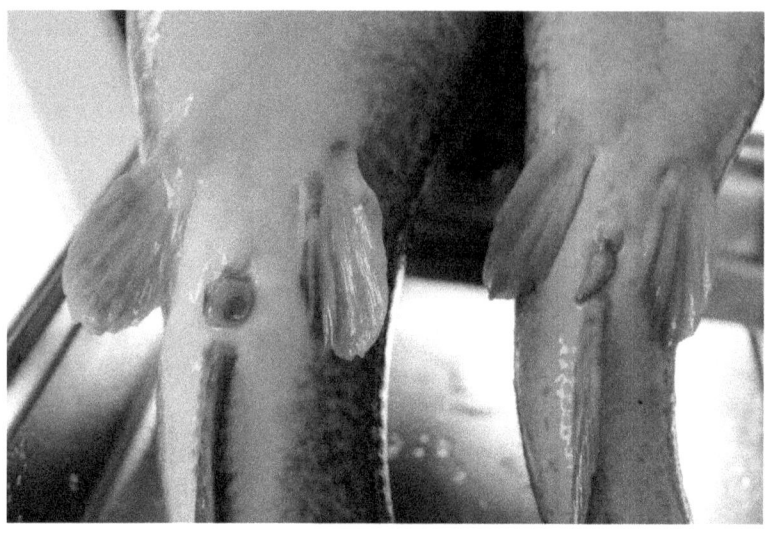

What Happens in the Broodstock Stage:

- Selection: The best-performing fish from each generation are selected as broodstock for the next breeding cycle. These fish need to be healthy, disease-free, and of optimal size.

- Breeding Cycle: The breeding cycle typically begins again when water temperatures rise, signalling the start of a new spawning season.

- Management: Broodstock require special care to ensure they remain healthy and productive. They need to be fed a nutrient-rich diet to maintain their reproductive capabilities.

Considerations for Farmers:

- Long-Term Care: Proper management of broodstock is essential to maintaining the quality of the offspring. Regular monitoring, disease prevention, and optimal feeding are key.

- Separate Tanks or Ponds: Broodstock should be kept separate from other catfish to ensure they are not overcrowded or stressed before the spawning season.

The life cycle of catfish—spanning from spawning, hatching, and growing through the juvenile fingerling stage, to reaching full adulthood—requires a careful and attentive management process. Each stage of development has its own unique requirements, whether it's feeding, water quality, or protection from disease and predators. As a catfish farmer, understanding these stages and knowing how to meet the needs of your stock at each phase will ensure a productive, healthy, and profitable farm. By providing the right environment, diet, and care at every stage of the catfish life cycle, you can maximize growth, minimize losses, and develop a thriving catfish farming operation.

1.3 Choosing the Right Breed for Your Farm

When starting a catfish farm, one of the most important decisions you'll make is selecting the right breed of catfish to farm. Not all catfish are the same—different species have unique characteristics that can significantly impact your farming success. Choosing the right breed will depend on your goals, local climate, market demand, and the resources you have available. Here's an explanation to help you make the best decision for your catfish farm:

Factors to Consider When Choosing a Catfish Breed

1. Climate and Water Conditions

Catfish thrive in a variety of climates, but some species are more tolerant of temperature fluctuations than others. Consider your local weather and the water conditions you can provide, such as:

- Temperature Range: Some catfish species prefer warmer water, while others can handle cooler temperatures. Ensure that the species you choose can thrive in your local environment or in controlled aquaculture conditions.

- Water Quality: Some catfish are more sensitive to water quality, particularly oxygen levels and pH. Make sure your farm can maintain optimal water conditions for the species you select.

2. Growth Rate and Size

Different catfish breeds have varying growth rates, which will affect how quickly your fish reach market size. A faster-growing breed can shorten the time to harvest, but some slower-growing species may offer higher-quality meat. Ask yourself:

- How quickly do you want your catfish to reach market size? Fast-growing species may allow for more frequent harvesting.

- What size of fish is in demand in your market? Some species grow larger than others, so it's essential to know what size fish your customers expect.

3. Market Demand and Local Preferences

Understanding what your target market wants is key to running a profitable catfish farm. Some species are more popular due to their flavour, texture, or ease of preparation. Consider:

- What type of catfish is popular in your region? Some markets may prefer a specific breed due to taste, availability, or cultural preferences.

- Are there specific requirements for export or high-end markets? Premium species or breeds with certain qualities may fetch higher prices but might also require more care and investment.

4. Disease Resistance and Hardiness

Different species of catfish have varying levels of resistance to diseases and environmental stressors. Hardier species may be easier to farm, especially for beginners. Consider:

- Are you prepared to deal with common fish diseases? If not, choose a species known for its resistance to diseases and stress.

- What is the mortality rate of the species you're considering? Hardier species with lower mortality rates can increase your chances of success.

5. Feed Efficiency

The cost of feed is one of the biggest expenses in catfish farming. Some species are more efficient at converting feed into body mass, meaning they grow faster with less food. Consider:

- How efficient is the breed in converting feed into weight? More feed-efficient species will help you reduce costs and improve profit margins.

- What kind of feed does the species require? Some breeds may require specialized feed, which could be more expensive or difficult to source.

How to Select the Best Breed for Your Farm

When choosing the best breed for your catfish farm, follow these steps:

- Evaluate Your Environment: Consider your local climate, water availability, and farming infrastructure to determine which breeds are best suited to your region.

- Assess Market Needs: Look at local and regional market trends to understand which species of catfish are in demand. Are restaurants, grocery stores, or local buyers asking for a specific breed?

- Consider Your Resources: If you have access to high-quality feed, you might opt for a breed that grows quickly. If water quality or resources are limited, choose a hardier species like African catfish.

- Plan for the Future: Some species may be more labour-intensive or require more time to grow. Decide if you want a faster-growing breed or are willing to invest more time and effort for higher profits from larger or specialty fish.

Choosing the right breed for your catfish farm is a critical step that can determine your farm's success. By understanding the climate, market demand, feed efficiency, and hardiness of different catfish species, you can select the breed that aligns with your farming goals and resources. Whether you're aiming for fast-growing, hardy fish like channel catfish or focusing on niche markets with larger species like blue or flathead catfish, your choice will set the foundation for a profitable and sustainable operation.

CHAPTER 2

PLANNING YOUR CATFISH FARM

2.1 Setting Goals for Your Catfish Farm

Before diving into catfish farming, it's essential to have a clear set of goals. Setting goals not only provides you with a roadmap for your business but also helps measure your progress, identify challenges, and plan for future growth. Whether you're starting small or aiming to build a large commercial operation, well-defined goals will ensure your catfish farm stays on track for success. Here's an in-depth guide on how to set meaningful goals for your catfish farm:

Why Goal Setting is Important for Your Catfish Farm

Just like any other business, catfish farming requires a well-thought-out plan. Without goals, it's easy to get lost in the day-to-day tasks without focusing on the bigger picture. Setting specific, achievable goals helps to:

- Provide Direction: Knowing what you want to achieve gives your farm a clear direction. It will guide your decisions, from selecting the right breed of catfish to choosing equipment, feed, and water systems.

- Measure Progress: With clear goals, you can track your farm's growth and success over time. If your goal is to produce a certain number of catfish by the end of the year, you can measure your progress month by month.

- Motivate and Focus: Having clear objectives keeps you and your team motivated. It reminds you why you started farming and encourages you to stay focused, even when challenges arise.

- Identify Resources and Needs: Goals help you assess what you'll need to succeed—whether it's funding, equipment, expertise, or labour. It allows you to prioritize and allocate resources efficiently.

- Adapt and Grow: Goals aren't static. As your farm evolves, you can adjust your objectives to meet new demands or take advantage of new opportunities, such as expanding your market or improving productivity.

Types of Goals for Your Catfish Farm

To succeed in catfish farming, your goals should be specific, realistic, and tailored to your unique situation. Goals can be broken down into different categories, helping you cover every aspect of the business:

1. Production Goals

Production goals revolve around how much catfish you plan to grow, harvest, and sell. These goals will depend on the size of your farm, the species you are raising, and your local market demand.

- Example 1: "Harvest and sell 10,000 pounds of catfish by the end of the year."

- Example 2: "Increase the survival rate of juvenile catfish by 15% over the next six months."

These goals help you focus on the operational aspects of farming, such as managing the growth cycle, improving water quality, feeding efficiently, and preventing diseases.

2. Financial Goals

Every business needs financial objectives to stay profitable. Setting financial goals for your farm will ensure you are keeping costs in check, generating enough revenue, and achieving your desired level of profit.

- Example 1: "Generate $50,000 in revenue by the end of the first year."
- Example 2: "Reduce feed costs by 10% within six months by improving feed conversion efficiency."

By keeping track of your income and expenses, you can make sure your farm remains financially healthy. Financial goals also encourage smart budgeting and resource allocation.

3. Expansion Goals

As your farm grows, you may want to set goals that focus on expanding your operations. These could involve increasing the size of your pond or tank system, adding new breeds, or exploring new markets.

- Example 1: "Increase production capacity by adding 2 additional ponds within the next year."
- Example 2: "Begin raising a new species of catfish within 18 months to diversify offerings."

Expansion goals allow you to plan for the long-term future of your farm. They give you a vision of what your farm could become, helping you make decisions that support growth.

4. Marketing and Sales Goals

Your farm's success depends on your ability to sell your catfish. Marketing and sales goals focus on building a strong customer base and expanding your reach in the market.

- Example 1: "Secure three new local restaurant contracts within six months."
- Example 2: "Sell 90% of harvested fish within two weeks of harvest."

These goals help you focus on where and how to sell your fish, whether it's through direct sales to consumers, selling to wholesalers, or securing long-term contracts with restaurants or retailers.

5. Sustainability and Environmental Goals

Sustainability is becoming increasingly important in farming. By setting environmental goals, you can ensure that your catfish farm operates in an eco-friendly and responsible manner.

- Example 1: "Reduce water usage by 20% by implementing a recycling system within the next year."
- Example 2: "Switch to sustainable feed sources within two years."

These goals focus on reducing waste, conserving resources, and minimizing the environmental impact of your farm. They can also help you market your farm as a responsible, sustainable business.

How to Set SMART Goals for Your Catfish Farm

When setting goals for your catfish farm, it's useful to follow the SMART method. SMART goals are:

i. Specific: Your goals should be clear and well-defined. Vague goals won't give you a clear direction or make it easy to measure your progress.

- *Instead of*: "I want to farm more catfish."
- *Try*: "I want to increase production by 25% over the next year."

ii. Measurable: Set goals that you can measure with clear milestones or numbers.

- *Instead of*: "I want my farm to be profitable."
- *Try*: "I want to make $10,000 in profit by the end of the year."

iii. Achievable: Set goals that are challenging yet realistic. While aiming high is important, setting unattainable goals will lead to frustration.

- *Instead of*: "I want to triple production in 6 months."
- *Try*: "I want to increase production by 10% in the first six months."

iv. Relevant: Make sure your goals align with your overall business objectives and the resources you have.

- *Instead of*: "I want to sell my fish internationally."
- *Try*: "I want to expand sales within my local region."

v. Time-Bound: Set a deadline or time frame for achieving your goals.

- *Instead of*: "I want to increase revenue."
- *Try*: "I want to increase revenue by 15% within the next 12 months."

Examples of SMART Goals for Your Catfish Farm

To give you a practical idea, here are a few SMART goals that you might consider setting for your farm:

- Production Goal: "Raise and harvest 5,000 pounds of catfish by the end of the year, ensuring that 90% of the stock reaches market size."

- Financial Goal: "Reduce operational costs by 10% in six months by improving feed conversion efficiency and reducing energy consumption."

- Marketing Goal: "Increase local sales by 20% within one year by establishing contracts with two local restaurants and participating in three farmers' markets."

- Expansion Goal: "Add a second pond system and double production capacity within two years."

- Sustainability Goal: "Implement a waste recycling system to reduce pond water usage by 15% within the next year."

Aligning Your Goals with Your Vision

Before setting specific goals, it's essential to think about your overall vision for the catfish farm. Ask yourself:

- What do you want to achieve in the next 5 to 10 years? Do you envision a small-scale operation or a large commercial farm?

- What are your personal motivations? Is catfish farming primarily a business venture for profit, or is sustainability and environmental responsibility a key focus?

- How do you want to impact your community? Are you interested in providing fresh, local fish to your community or supporting food security initiatives?

Your answers to these questions will help shape the goals you set for your farm and guide your decisions as your business grows.

Setting clear, achievable goals is the foundation of a successful catfish farm. Whether your focus is on increasing production, maximizing profits, expanding your operations, or running an environmentally sustainable business, having the right goals in place ensures you stay on track and grow steadily. By taking the time to define SMART goals, you can build a roadmap for success, track your progress, and ultimately turn your vision for your catfish farm into a reality.

2.2 Selecting a Suitable Location for Your Catfish Farm

Choosing the right location for your catfish farm is one of the most crucial decisions you'll make. The location will significantly impact your farm's productivity, profitability, and sustainability. From the availability of water to the accessibility of markets, multiple factors must be considered to ensure you pick a spot that will support the success of your catfish farming operation. In this guide, we'll walk you through the key considerations when selecting a location for your catfish farm and explain how each one plays a role in your farm's long-term success.

Why Location Matters in Catfish Farming

The location of your catfish farm affects nearly every aspect of your operation, including:

- Water Supply: Catfish need a clean, reliable water source to thrive. The quality and quantity of water available at your farm location will determine the health and growth of your stock.

- Climate: The local climate influences water temperature, which can affect catfish growth rates, feeding behaviour, and reproduction.

- Land and Space: Adequate land is necessary to construct ponds or tanks, build infrastructure, and allow for future expansion.

- Accessibility: The ease of transporting feed, equipment, and harvested catfish to and from your farm will impact costs and efficiency.

- Regulations: Local regulations or zoning laws may affect where you can build your farm, how you manage water, and how you handle waste.

By taking all these factors into account, you can choose a location that provides an optimal environment for raising healthy, productive catfish.

Factors to Consider When Selecting a Location

1. Water Availability and Quality

Water is the lifeblood of any catfish farm. Ensuring that you have access to an abundant and reliable water supply is essential for maintaining healthy ponds or tanks. When evaluating a potential location, consider:

- Water Source: Your farm will require a consistent supply of fresh water from sources like rivers, lakes, underground wells, or rainwater. The quantity of water available should be enough to fill your ponds, maintain water levels, and allow for periodic water changes to keep the environment clean.

- Water Quality: The water should be free from pollutants, harmful chemicals, or toxins that could harm your fish. Test the water for parameters such as pH levels, dissolved oxygen content, and ammonia concentrations. Clean, well-oxygenated water is critical to catfish health and growth.

- Water Management: You'll also need a drainage system to remove excess water during cleaning or maintenance.

Ensure your location allows for proper water disposal that meets environmental regulations.

2. Land Area and Soil Conditions

The amount of land you need will depend on the scale of your operation. However, it's essential to choose a site that provides enough space for your ponds, tanks, or raceways, and leaves room for expansion. When evaluating the land:

- Size of the Land: Ensure you have enough land to build multiple ponds or tank systems. As your farm grows, you may want to expand, so it's a good idea to leave room for future development.

- Soil Type: If you're building earthen ponds, the type of soil at the location matters. Clay soils are ideal for pond construction because they retain water well, while sandy soils may drain water too quickly, leading to higher water usage and costs. Have the soil tested to ensure it's suitable for pond construction.

- Topography: A relatively flat or gently sloping area is preferable for building ponds, as this makes it easier to manage water flow and drainage. Avoid areas prone to flooding, as this can damage your ponds and lead to fish escaping.

3. Climate and Temperature

Catfish farming requires a stable environment where water temperature stays within a range that promotes optimal growth. The climate of your farm's location will have a direct impact on the water temperature, which in turn affects feeding, reproduction, and overall health.

- Ideal Temperature Range: Catfish thrive in water temperatures between 75°F and 85°F (24°C to 29°C). If your location has extreme temperatures outside this range, you may need to invest in temperature control systems, which can be costly.

- Seasonal Variations: Consider the seasonal weather patterns in your area. Extremely cold winters or scorching hot summers may require adjustments to your farming practices, such as using indoor tanks or shading systems for outdoor ponds.

- Rainfall and Water Levels: Areas with consistent rainfall may provide a natural source of water for your ponds. However, excessive rain can cause ponds to overflow, so ensure proper drainage systems are in place.

4. Accessibility and Proximity to Markets

Your catfish farm needs to be accessible for the delivery of supplies (such as feed and equipment) and the transport of harvested fish to buyers. Consider the following:

- Road Access: A location with good road infrastructure is essential for transporting fish, which are perishable and need to be delivered quickly to market. Farms located in remote areas with poor roads may face higher transportation costs and logistical challenges.

- Proximity to Markets: Being close to your target market reduces transportation time and costs. If you plan to sell your catfish to restaurants, grocery stores, or directly to consumers, choosing a location near urban centres can be advantageous.

- Suppliers and Services: You'll also want to be near suppliers of feed, equipment, and other farm necessities. Being too far from these suppliers can lead to delays and increased expenses.

5. Legal and Environmental Regulations

Before settling on a location, it's important to research any local regulations, zoning laws, and environmental guidelines that apply to fish farming. Regulatory considerations include:

- Zoning: Ensure that the land is zoned for agricultural or aquaculture use. Some areas may have restrictions on commercial farming operations.

- Permits and Licenses: Depending on your location, you may need specific permits to operate a catfish farm, use water from nearby sources, or discharge water from your ponds. Make sure to understand the legal requirements in your region.

- Environmental Impact: Fish farming can have environmental impacts, such as water usage and waste management. Ensure that your farm complies with environmental regulations to avoid penalties and negative consequences for the local ecosystem.

6. Infrastructure and Utilities

A catfish farm requires basic infrastructure to operate efficiently. Look for a location that provides:

- Electricity: You'll need a reliable power supply to run pumps, aerators, and other equipment used in managing ponds or tanks. In remote areas, consider alternative energy sources like solar power.

- Roads and Transportation: Easy access to main roads allows for the efficient delivery of supplies and the transportation of harvested fish.

- Proximity to Skilled Labour: If you plan to hire workers, it's beneficial to locate your farm in an area where skilled labour is readily available.

Balancing Cost with Location Benefits

While you want a location that offers all the above advantages, it's essential to balance your choices with the cost of land, utilities, and infrastructure development. Rural areas might offer more affordable land, but the cost of transporting fish and supplies may be higher. On the other hand, more accessible or urban locations might come with higher land prices but reduce operational costs due to better infrastructure and proximity to markets.

Ultimately, the key is to find a location that aligns with your budget while meeting the essential criteria for successful catfish farming.

Example: What Makes a Great Location?

Let's imagine you're evaluating two potential locations for your catfish farm:

1. Location A is a 5-acre plot near a river in a rural area. The water supply is abundant and clean, and the land is relatively flat, making it easy to construct ponds.

However, the farm is 50 miles away from the nearest market and supplier, which could increase transportation costs.

2. Location B is a 3-acre plot closer to a small town, with access to good roads and nearby markets. There's a reliable water supply from an underground well, but the land is more expensive, and the soil will require modification to build efficient ponds.

Both locations have their pros and cons. Location A offers cheaper land and a natural water source, but Location B reduces transportation costs and puts you closer to your market. The best choice depends on your priorities, whether you value lower land costs or better accessibility.

Selecting the right location for your catfish farm is a crucial step that can make or break your farming venture. A site with a reliable water source, suitable land, favourable climate, and good accessibility will create the foundation for a thriving catfish operation. While no location will be perfect, balancing these factors with your budget and long-term goals will help you choose a site that sets you up for success. By carefully considering water availability, soil conditions, climate, and access to markets, you'll be well on your way to starting a catfish farm that's both productive and profitable.

2.3 Legal and Regulatory Requirements for Catfish

Farming

Starting a catfish farm is not just about choosing the right location or breed of fish—it also requires a solid understanding of the legal and regulatory framework that governs aquaculture in your area. This ensures that your farm operates within the law, prevents costly fines or penalties, and helps you establish a sustainable business that's in harmony with the environment and local community.

This section will explain the key legal and regulatory aspects you need to consider when starting a catfish farm, including permits, licenses, water usage regulations, environmental compliance, and labour laws. By understanding these requirements early, you can avoid potential legal issues and run a successful, compliant farm.

Why Legal Compliance is Important

Operating within the legal framework has multiple benefits:

- Avoid Penalties: Ignoring regulations can result in hefty fines, closure of your farm, or even legal action.

- Environmental Responsibility: Regulatory compliance ensures that your farm does not harm local ecosystems, protecting water sources and wildlife.

- Building Trust: By adhering to regulations, you build trust with customers, suppliers, and local authorities, which is vital for long-term success.

- Smoother Operations: Legal clarity allows you to focus on managing and growing your farm without worrying about unforeseen legal hurdles.

Legal and Regulatory Areas for Catfish Farming

1. Business Registration and Licensing

Before you can start your catfish farm, you'll need to officially register your business and obtain the necessary licenses to operate. This step ensures that your farm is recognized by the government as a legal entity.

- Business Registration: Depending on your country or region, you may need to register your farm as a sole proprietorship, partnership, or limited liability company (LLC). This step legally establishes your business and allows you to operate under a recognized business name.

- Operating License: Aquaculture, including catfish farming, often requires a specific operating license. This license certifies that your farm complies with the necessary farming and environmental regulations. Check with your local agricultural or fisheries department to determine what licenses are required for catfish farming in your area.

2. Aquaculture Permits

Since catfish farming involves the cultivation of fish, you may need to apply for an aquaculture permit that governs how you manage your ponds, water use, and fish stocks.

- Aquaculture Permit: This permit is typically issued by local fisheries or environmental agencies. It regulates the size of your farm, the type of fish you can raise, and your farming practices. The permit may include restrictions on the number of catfish you can farm and guidelines on how to prevent diseases and contamination.

- Endangered Species and Biodiversity Protection: If your farm is located near natural water bodies, there may be restrictions on introducing non-native or invasive species of fish. It's essential to ensure that your farming activities do not negatively affect local biodiversity or harm endangered species.

3. Water Usage and Rights

Water is one of the most critical resources for catfish farming. Governments often regulate the use of water to ensure sustainability and fair distribution. This means that you will need to comply with local water rights and usage regulations.

- Water Use Permit: In many regions, you need a permit to draw water from rivers, lakes, or underground wells for use on your farm. This permit ensures that your water usage does not negatively affect other users or deplete water resources.

- Wastewater Management: You will also need to manage the discharge of wastewater from your farm. Water from your ponds may contain organic matter, fish waste, and chemicals that can pollute local waterways. Local authorities may require you to treat wastewater before releasing it into the environment, or you may need to recycle it for use on your farm.

4. Environmental Regulations

Aquaculture can have an impact on the environment if not properly managed. Governments set environmental regulations to minimize these impacts and ensure that farming practices are sustainable. Compliance with environmental regulations is not only good for the planet but also essential for the long-term success of your catfish farm.

- Environmental Impact Assessment (EIA): In some areas, you may be required to conduct an Environmental

Impact Assessment before starting your farm. This assessment evaluates the potential environmental effects of your farming activities, such as habitat destruction, water pollution, or effects on local wildlife. Based on the findings, authorities may impose specific conditions or restrictions to minimize your farm's environmental footprint.

- Sustainable Farming Practices: Many governments encourage sustainable farming practices that protect the environment. This could include guidelines on pond construction, the use of organic or chemical-free feed, proper waste disposal, and measures to prevent water contamination.

5. Health and Safety Regulations

Catfish farms must comply with health and safety regulations to ensure the safety of farm workers, fish stock, and consumers. These rules help prevent the spread of diseases and ensure the overall well-being of everyone involved in the farming process.

- Fish Health Management: You must follow regulations on fish health management, which may include routine inspections by government agencies. These inspections ensure that your fish are free from diseases that could spread to other farms or harm consumers. Vaccinations, quarantining new fish, and disease monitoring are essential practices for maintaining a healthy stock.

- Biosecurity Protocols: Biosecurity measures prevent the introduction and spread of diseases on your farm. Regulations may require specific protocols, such as controlling farm access, disinfecting equipment, and managing water flow to prevent contamination.

- Food Safety: If you are processing or selling catfish for human consumption, you'll need to comply with food safety regulations. This ensures that the fish you sell is safe for consumers and meets hygiene standards. In some areas, this may include certifications for handling, packaging, and storing fish.

6. Zoning and Land Use Regulations

The location of your catfish farm may be subject to zoning laws, which govern how land can be used in a particular area. These regulations ensure that farming activities are compatible with the surrounding community and environment.

- Zoning Laws: Before purchasing land for your farm, check local zoning regulations to ensure that aquaculture is permitted on the property. Some areas may be zoned for residential, commercial, or industrial use, which could limit your ability to start a farm.

- Building Permits: If you are constructing ponds, tanks, or other infrastructure, you may need building permits from local authorities. This ensures that all structures are safe and meet local building codes.

7. Labour Laws

If you plan to hire workers for your catfish farm, you'll need to comply with local labour laws. These laws govern wages, working conditions, and worker safety.

- Minimum Wage and Fair Employment: Ensure that your farm pays employees the legal minimum wage and adheres to fair labour practices. This includes providing proper contracts and ensuring that working hours comply with local laws.

- Occupational Safety: Catfish farming can involve manual labour, heavy machinery, and handling live animals. You are responsible for creating a safe work environment by providing training, safety gear, and complying with regulations on workplace safety.

8. Taxation and Financial Regulations

Like any other business, your catfish farm will be subject to taxation. This includes income tax, property tax, and possibly special taxes related to farming activities.

- Tax Registration: Ensure that your business is registered for tax purposes. You'll need to keep detailed records of income and expenses, and file taxes regularly according to your local tax laws.

- Subsidies and Grants: In some regions, governments offer subsidies or grants to support the development of the aquaculture industry. These financial incentives can help offset start up costs or support sustainable farming practices. Research available programs in your area to see if you qualify.

<u>Navigating Legal and Regulatory Requirements: Practical Tips</u>

- Consult Local Authorities: Contact your local agricultural or fisheries department to find out about specific legal requirements in your area. They can provide guidance on permits, licensing, and regulations.

- Seek Professional Help: If you're unsure about legal matters, consider hiring a lawyer or consultant with experience in aquaculture. They can help you navigate the complexities of regulations and ensure that your farm is fully compliant.

- Stay Informed: Regulations change over time, so it's essential to stay updated on any new laws or policies that may affect your farm. Regularly check with local authorities or industry associations to ensure you remain compliant.

Starting a catfish farm involves more than just managing ponds and fish; it also requires a thorough understanding of the legal and regulatory landscape. By obtaining the necessary permits, complying with environmental and health regulations, and adhering to labour laws, you can establish a farm that operates smoothly and avoids costly legal issues. Being proactive about legal compliance not only protects your farm from penalties but also builds a strong foundation for long-term success. Understanding and following these regulations will allow you to focus on what matters most: raising healthy, profitable catfish.

CHAPTER 3

BUILDING THE INFRASTRUCTURE

3.1 Types of Farming Systems in Catfish Farming

When starting a catfish farm, one of the first decisions you'll need to make is choosing the right farming system. The farming system you use will impact your farm's layout, management practices, costs, and the overall success of your catfish production. Understanding the different types of farming systems is crucial for selecting the one that best fits your goals, resources, and location. This section will explore the primary catfish farming systems, their advantages, disadvantages, and how to choose the one that suits your needs.

Overview of Catfish Farming Systems

Catfish farming systems can be categorized based on how fish are raised and the scale of operation. The three main types are:

- Extensive Farming Systems: Low-density, natural food-dependent ponds.

- Semi-Intensive Farming Systems: Moderate-density ponds with supplemental feeding.

- Intensive Farming Systems: High-density ponds or tanks with complete control over feeding and water quality.

Each system has its unique characteristics, and the choice will depend on factors such as the size of your farm, available water resources, budget, and the level of management you can provide.

1. Extensive Farming Systems

An extensive farming system is the most basic and traditional method of catfish farming. It relies heavily on natural resources, including sunlight, water, and the pond's natural ecosystem, to support the growth of catfish. In this system, fish density is kept low, and the primary source of food for the fish is naturally occurring plankton and aquatic plants.

How It Works

- Catfish are stocked in a large pond, usually at a low density (fewer fish per unit of water).
- The pond relies on natural food sources like algae, plankton, and insects.
- There is minimal human intervention; supplemental feed is rarely used.
- Water quality is naturally maintained through the ecosystem's balance.

Advantages

- Low-Cost Operation: Since the system uses natural food and water without much human intervention, the operational costs are low.
- Eco-Friendly: This system mimics a natural environment, minimizing the environmental impact and maintaining ecological balance.
- Low Labour Requirement: Extensive systems require less daily management and labour since the fish feed on natural resources.

Disadvantages

- Low Productivity: The growth rate of catfish is slower due to the limited availability of food, leading to lower yields.

- Vulnerability to Environmental Factors: Changes in weather, water temperature, or water quality can significantly impact fish health and growth.

- Limited Control: Since there's no control over the feeding and water quality, it's difficult to manage the fish's growth, health, and disease prevention.

Who Should Use This System?

- Farmers with large land areas and access to natural water sources.

- Those looking for a low-cost, low-maintenance way to farm catfish.

- Hobbyists or small-scale farmers not focused on high production volumes.

2. Semi-Intensive Farming Systems

Semi-intensive farming systems are a step up from extensive systems. This system combines natural food production with supplemental feeding and moderate control over the environment. Farmers provide additional feed to ensure that the catfish grow faster and reach market size more quickly.

How It Works

- Catfish are stocked at a moderate density in ponds.

- Fish are fed a combination of natural food (plankton, insects) and supplemental feeds such as commercial pellets or homemade feed.

- Farmers monitor water quality and occasionally intervene to maintain optimal conditions.

- Some water management practices, such as aeration or water exchange, may be used to maintain the health of the fish.

Advantages

- Higher Productivity: With supplemental feeding, catfish grow faster and reach a larger size, resulting in higher yields compared to extensive systems.

- Moderate Costs: Although there are additional costs for feed and water management, the overall cost is still relatively low.
- Flexible Management: Farmers have more control over the fish's diet, health, and growth, making it easier to manage disease and ensure good water quality.

Disadvantages

- Increased Labour: Semi-intensive systems require more hands-on management, including feeding, monitoring water quality, and performing routine maintenance.
- Environmental Impact: The use of supplemental feeds and occasional water exchanges can contribute to nutrient build up in the water, leading to environmental concerns such as water pollution or algae blooms.
- Moderate Investment: You'll need to invest in supplemental feeds, basic water management tools (aerators, pumps), and monitoring equipment.

Who Should Use This System?

- Farmers who want to balance productivity with manageable costs and labour.
- Those who have some experience with fish farming or are willing to invest in supplemental feeding and water quality management.
- Farmers aiming for a moderate-scale operation with higher yields than an extensive system.

3. Intensive Farming Systems

Intensive farming systems are highly controlled environments where catfish are raised at high densities, usually in tanks, raceways, or ponds with advanced water management systems. In this system, all aspects of the fish's environment—feeding, water quality, oxygen levels, and waste management—are closely monitored and controlled to maximize growth and production.

How It Works

- Catfish are stocked at high densities in either large tanks, raceways, or highly managed ponds.
- Fish are fed entirely on commercial feeds or specially formulated diets.
- Advanced systems like aeration, filtration, and water recycling are used to maintain optimal water quality and oxygen levels.
- Farmers monitor the fish's health, growth, and water conditions regularly, making adjustments as needed.

Advantages

- High Productivity: Intensive systems produce the highest yield per unit area, allowing farmers to grow more fish in a smaller space.
- Full Control: Farmers have complete control over feeding, water quality, oxygen levels, and disease management, leading to faster growth and healthier fish.
- Year-Round Production: With the ability to control water temperature and quality, intensive systems can support year-round production, independent of seasonal changes.

Disadvantages

- High Costs: Intensive systems require significant investment in infrastructure (tanks, water pumps, filtration systems) and commercial feed.
- Complex Management: These systems require advanced knowledge and experience in aquaculture, as well as constant monitoring and intervention.
- Environmental Concerns: The intensive use of feed and water resources can lead to waste build up and environmental pollution if not managed properly.

Who Should Use This System?

- Farmers looking to maximize production on a small or medium-sized farm.

- Experienced farmers or those with the resources to hire trained staff for daily management.
- Those with access to capital for initial investment and who are focused on high yields and profitability.

Other Systems to Consider

In addition to the main systems mentioned, there are some other catfish farming techniques you might encounter or consider incorporating into your operation:

- Cage Farming: Fish are raised in cages or nets placed in natural water bodies like rivers or lakes. This system allows farmers to use existing water resources, but it offers less control over water quality and disease management.
- Recirculating Aquaculture Systems (RAS): This is a highly intensive system where water is constantly filtered and recirculated, making it one of the most sustainable methods. RAS allows you to farm catfish indoors or in urban settings, but it requires substantial upfront investment in filtration technology and constant management.

How to Choose the Right System for Your Farm

When deciding on the best farming system for your catfish operation, consider the following factors:

- Scale of Production: If you're aiming for large-scale production, an intensive system might be the best choice, while smaller operations may thrive with semi-intensive or extensive systems.
- Budget: Extensive systems are the most affordable but yield lower returns. Intensive systems offer higher returns but require significant investment.
- Land and Water Resources: Consider the availability of land and water. Extensive systems require large ponds and abundant natural water sources, while intensive systems can work with smaller, more controlled environments.

- Labour and Management: Evaluate how much time and labour you can dedicate. Intensive systems require constant monitoring, while extensive systems are more hands-off.

- Environmental Impact: If you prioritize sustainability, semi-intensive or extensive systems might be more appealing, as intensive farming can have a greater environmental footprint.

The type of farming system you choose will greatly influence your catfish farm's productivity, costs, and environmental impact. Extensive systems are low-cost and eco-friendly but have lower yields, while semi-intensive and intensive systems offer higher production but require more investment and management. By carefully considering your farm's goals, available resources, and level of expertise, you can choose the farming system that best aligns with your vision for success. Each system has its pros and cons, so it's essential to weigh these factors before making a decision.

3.2 Pond Construction and Design for Catfish Farming

Constructing and designing a pond for catfish farming is one of the most critical steps in establishing a successful farm. The design and construction of your pond will impact the growth, health, and overall productivity of your catfish, as well as how easily you can manage the farm. A well-built pond not only creates a healthy environment for the fish but also minimizes maintenance and operational costs in the long run. In this section, we will explore the essential aspects of pond construction and design, including site selection, pond size, shape, depth, and necessary infrastructure. Whether you are building a new pond or renovating an existing one, these principles will help ensure your catfish farm thrives.

1. Site Selection

Choosing the right location for your catfish pond is the first and most crucial decision. The site should meet the following criteria:

Considerations:

- Water Source: Ensure a reliable and sufficient source of water, whether from underground wells, rivers, or rainwater collection. The water quality should be good, with no contamination from chemicals, pollutants, or harmful microorganisms.

- Topography: The land should be gently sloping to facilitate water flow and drainage. Flat or low-lying areas can lead to poor drainage, while steep areas might cause erosion.

- Soil Type: The soil must be able to retain water. Clay or loamy soils are ideal, as they are less permeable and prevent water from leaking out of the pond. Sandy or porous soils may need lining or compaction to retain water.

- Proximity to Infrastructure: Consider access to electricity, roads, and transportation, as well as the proximity to markets for selling your fish.

- Flood Risk: Avoid areas prone to flooding, as this could damage the pond structure or wash away your fish stock.

2. Pond Size and Shape

The size and shape of your pond should align with your farming goals, land availability, and budget. The goal is to create a pond that is both efficient for fish growth and easy to manage.

Pond Size:

- Small-Scale Farms: Ponds can range from 500 to 1,000 square meters (5,000 to 10,000 square feet). Smaller ponds are easier to manage and are suitable for beginners or farmers with limited space.

- Medium to Large Farms: Larger ponds of 2,000 to 5,000 square meters (20,000 to 50,000 square feet) or more are used for commercial production. These allow for more fish and higher yields but require more management and resources.

Pond Shape:

- Rectangular: Most catfish ponds are rectangular because they are easier to manage, especially for feeding, water flow, and harvesting. Rectangular ponds also ensure better water circulation and oxygenation, which are vital for fish health.

- Irregular or Circular: These are less common but may be used in specific situations, such as adapting to the natural landscape. However, they can be more challenging to manage in terms of water flow and fish harvesting.

Design Tips:

- Avoid sharp corners in the pond's design, as these can trap fish and lead to uneven water flow.

- Ensure the pond is deep enough to allow for natural water temperature regulation and fish comfort, but not too deep to hinder oxygen diffusion or management.

3. Pond Depth

The depth of your catfish pond is another essential factor for maintaining a stable environment. The right depth will allow for proper water temperature regulation, fish comfort, and ease of management.

Recommended Depth:

- Optimal Depth: 1.5 to 2 meters (5 to 6.5 feet) is considered the ideal depth for a catfish pond. This depth allows for stable water temperatures and sufficient oxygenation while ensuring the pond doesn't become too shallow during dry seasons.

- Shallow Edges: Ponds should have gradually sloping edges, as these areas can be useful for fish feeding and easier for the farmer to monitor and manage.

Why Depth Matters:

- Temperature Control: Deeper ponds are more stable in terms of temperature, providing a comfortable environment for catfish to thrive. Shallow ponds can heat up too quickly in hot weather, leading to stress or disease outbreaks.

- Oxygen Levels: Adequate depth helps maintain better oxygen levels. Catfish are hardy but still require well-oxygenated water to stay healthy and grow. Deep ponds also prevent oxygen depletion, which can lead to fish death.

4. Water Supply and Drainage System

A reliable water supply and a well-designed drainage system are vital for maintaining water quality in your pond. The ability to refill and drain your pond easily helps you manage waste build up, water pollution, and fish health.

<u>Water Inlet System:</u>

- Water Source: The water used to fill your pond must be clean, fresh, and uncontaminated. It's best to filter the water at the inlet to prevent debris, pollutants, or predators from entering the pond.
- Inlet Design: Design the inlet in such a way that it introduces fresh water evenly across the pond. The inlet should be above the water level to avoid disturbing the pond's sediment and to allow aeration.

<u>Drainage System:</u>

- Outlet and Overflow: Every pond should have an outlet for draining water and an overflow system to prevent flooding during heavy rainfall. The drainage system should allow you to completely empty the pond when necessary for cleaning or harvesting.
- Sloping Design: The pond's bottom should slope gently toward the drain to facilitate easy removal of waste and water.

<u>Water Circulation:</u>

- Aeration: Installing aerators or water pumps can help improve water circulation, ensuring that oxygen is distributed evenly throughout the pond. Aeration also prevents the build up of toxic gases like ammonia, which can harm fish.

5. Pond Lining and Sealing

To prevent water leakage and maintain water quality, some ponds require lining or sealing, especially in areas where the soil is too sandy or porous to retain water naturally.

Types of Pond Liners:

- Clay Lining: Clay soil is naturally waterproof and is often compacted at the bottom and sides of the pond to prevent water loss. This is the most cost-effective method for sealing a pond.

- Plastic Liners: High-density polyethylene (HDPE) or PVC liners are often used in ponds with sandy soil or for added durability. These liners prevent water from seeping out and offer a long-lasting solution.

- Concrete Lining: Some large commercial farms opt for concrete-lined ponds, which offer excellent water retention and durability but are more expensive to build.

Sealing Tips:

- If using a clay liner, ensure that the clay is well-compacted to prevent cracks or water seepage.

- For plastic liners, ensure they are UV-resistant and properly anchored to the pond's edges to prevent damage over time.

6. Pond Infrastructure

Along with the basic design elements, there are additional infrastructures you may need to install for the effective management of your catfish pond.

Fencing:

- Purpose: A fence around the pond prevents predators (such as birds or animals) from accessing the pond and damaging your fish stock. Fencing also helps keep out trespassers and prevents accidents, particularly in areas with children or livestock.

Feeding Platforms:

- Feeding Areas: Designate specific areas of the pond for feeding to make it easier to monitor how much food your fish are consuming. These areas can be platforms or marked zones where you can observe the fish during feeding times.

Harvesting Infrastructure:

- Harvest Sumps or Harvest Basins: Including a harvest basin or sump at the bottom of the pond (near the drain) simplifies the harvesting process. When draining the pond, the fish naturally flow toward the sump, where they can be collected more easily.

- Pipes and Nets: For semi-intensive or intensive farms, using pipes or netting systems can make fish harvesting more efficient and reduce labour costs.

7. Pond Maintenance and Monitoring

Once your pond is constructed, regular maintenance and monitoring are essential to ensure that it functions optimally.

Regular Maintenance Tasks:

- Water Quality Testing: Regularly test the water for pH, ammonia, nitrate levels, and oxygen content to ensure it's within the optimal range for catfish.

- Sediment Removal: Over time, organic matter, such as fish waste and uneaten food, will accumulate at the bottom of the pond. Periodic cleaning and sediment removal help maintain water quality.

- Aerator Maintenance: If you use aerators, regularly check and maintain them to ensure they are functioning correctly and providing adequate oxygen levels.

Pond construction and design play a foundational role in the success of a catfish farm. A well-designed pond not only creates a healthy environment for your fish but also makes management tasks like feeding, harvesting, and maintaining water quality easier and more efficient. By carefully considering factors such as site selection, pond size and depth, water supply, and drainage systems, you can create a productive and sustainable catfish farming operation. Choosing the right materials for pond lining, investing in proper infrastructure, and staying on top of maintenance will ensure that your pond operates smoothly, maximizing your farm's productivity while minimizing potential challenges.

3.3 Setting Up Tanks and Equipment for Catfish Farming

In addition to pond systems, many catfish farmers use tanks for farming, especially when space is limited or when more controlled environments are preferred. Tanks offer flexibility, easier management, and better monitoring of fish health and water quality. Setting up tanks for catfish farming involves selecting the right type of tanks, setting up the necessary equipment, and creating an environment that promotes healthy growth and high yield. Here, we will explore the key aspects of setting up tanks and the essential equipment needed to run a successful tank-based catfish farm.

1. Types of Tanks for Catfish Farming

Tanks for catfish farming come in different materials and designs, each with its advantages and disadvantages. The choice of tank depends on the size of your farm, budget, and available space.

Types of Tanks:

i. Plastic or Fiberglass Tanks: These are popular choices for small- to medium-scale catfish farming due to their durability, affordability, and ease of cleaning. Plastic tanks are lightweight and come in various sizes, making them ideal for farms with limited space.

ii. Concrete Tanks: Concrete tanks are sturdy, long-lasting, and suitable for larger, more permanent farms. They are more expensive to build but offer excellent durability and can accommodate larger volumes of fish.

iii. Metal Tanks: Although not as common, metal tanks can be used, provided they are coated with a non-toxic lining to prevent rust and contamination of the water.

iv. Aquaponics or Recirculating Systems: In more advanced or sustainable setups, tanks may be part of an aquaponic system where plants are grown using the nutrients from fish waste. Recirculating systems constantly filter and reuse water, making them eco-friendly and efficient.

Tank Size:

The size of the tanks you choose will depend on the number of catfish you intend to raise and your available space.

- Small-Scale Farming: For beginners or those with limited space, tanks of 500 to 2,000 litres (132 to 528 gallons) are suitable. These tanks are easier to manage and can accommodate a decent stock of catfish for small-scale production.

- Large-Scale Farming: Larger tanks, ranging from 5,000 litres (1,320 gallons) to 10,000 litres (2,640 gallons) or more, are often used in commercial farming. These tanks can hold a greater number of fish but require more monitoring and maintenance.

2. Tank Location and Layout

The placement of your tanks is crucial for the farm's efficiency, especially when it comes to feeding, cleaning, and monitoring. Tanks can be placed indoors, in greenhouses, or outdoors, depending on your setup.

Considerations:

- Accessibility: Place your tanks in a location where you can easily access them for feeding, water changes, and monitoring. Ensure there's enough space between tanks for easy movement.

- Proximity to Water Source: Tanks should be close to a reliable water source for easy filling and draining. Whether you're using municipal water, well water, or rainwater, having a convenient supply will simplify daily operations.

- Drainage: Make sure your tank setup includes a proper drainage system to allow for water removal and waste management. Tanks should be slightly elevated to enable gravity-assisted draining.

3. Essential Equipment for Tank-Based Catfish Farming

Running a tank-based catfish farm requires specialized equipment to maintain optimal water quality, feed efficiency, and overall fish health. Below are the essential pieces of equipment needed:

i. Water Filtration System: Maintaining clean water is vital for the health and growth of catfish, especially in a tank environment where water is not naturally refreshed, as in a pond.

- Mechanical Filters: These remove physical particles such as uneaten food, waste, and debris. A fine mesh or sponge filter is commonly used for this purpose.

- Biological Filters: These promote the growth of beneficial bacteria that break down ammonia (from fish waste) into less harmful nitrates. A biofilter is crucial for maintaining a stable and safe environment for the fish.

- Chemical Filters: These use activated carbon or other materials to remove chemicals, toxins, and odours from the water.

ii. Aeration System: Catfish require oxygen-rich water to thrive. In tanks, natural aeration is limited, so an aeration system is necessary to keep oxygen levels high.

- Air Pumps: These pump air into the tank, ensuring that oxygen is distributed throughout the water. Air stones or diffusers are attached to the pump to break up air into small bubbles, maximizing oxygen absorption.

- Aerators: For larger tanks, aerators may be used to increase oxygen levels. They stir the water, allowing more oxygen to dissolve into the tank.

Adequate aeration prevents oxygen depletion, which can lead to fish stress or death, especially in densely stocked tanks.

iii. Water Heaters and Thermometers: Catfish are warm-water fish, thriving best in water temperatures between 25°C to 30°C (77°F to 86°F). In colder climates or during winter months, maintaining this temperature range is essential for their growth and survival.

- Water Heaters: Submersible heaters are commonly used to regulate the water temperature in tanks. These heaters come with adjustable thermostats that allow you to set the desired temperature.

- Thermometers: Install a water thermometer in each tank to monitor temperature levels. Regular checks will help you prevent sudden temperature fluctuations, which can stress or kill the fish.

iv. Feeding System: Proper feeding is crucial for the healthy growth of catfish, and tanks offer a controlled environment where feeding can be easily managed.

- Manual Feeding: Many small-scale farmers feed their fish manually using floating or sinking pellets. This method allows you to monitor the fish's feeding behaviour and adjust quantities accordingly.

- Automatic Feeders: For larger farms or when you are away, automatic feeders can dispense the right amount of food at set intervals. This ensures the fish are fed consistently and prevents overfeeding, which can pollute the water.

v. Tank Cleaning Tools: Regular tank cleaning is essential for preventing the build up of waste, uneaten food, and algae, which can compromise water quality and lead to disease.

- Siphon Hose: A siphon hose allows you to remove debris, waste, and excess food from the bottom of the tank without needing to empty the tank completely.

- Scrub Brushes: Use brushes or sponges to clean the tank walls and equipment, such as filters and aerators, to prevent algae growth.

- Water Testing Kits: These kits help you regularly check water parameters such as pH, ammonia, nitrite, and nitrate levels. Keeping these levels in check ensures the water remains healthy for your catfish.

4. Monitoring and Maintenance of Tank Systems

Setting up tanks is just the beginning; maintaining them requires regular monitoring and care to ensure a healthy environment for your fish.

i. Water Quality Monitoring: Constant monitoring of water quality is critical in tank-based systems. Regularly test for:

- pH Levels: The ideal pH for catfish is slightly acidic to neutral, ranging between 6.5 and 7.5.

- Ammonia and Nitrite: High levels of ammonia or nitrite can be toxic to fish. A biological filtration system helps manage these toxins, but regular testing is still necessary.

- Oxygen Levels: Use a dissolved oxygen meter to ensure that the oxygen levels remain optimal, especially in densely stocked tanks.

ii. Routine Water Changes: Even with filtration systems, regular partial water changes are necessary to remove dissolved waste and maintain water quality. Aim to change 10-20% of the water in each tank every week, depending on the stocking density and size of the tank.

iii. Equipment Maintenance: Regularly inspect and maintain your equipment, such as filters, heaters, and aerators, to ensure they are functioning correctly. Clean filters and replace parts as needed to avoid breakdowns that could harm your fish.

Setting up tanks and equipment for catfish farming provides a controlled environment where you can closely monitor and manage the growth of your fish. With the right tank size, location, and equipment such as filtration systems, aerators, and water heaters, you can create optimal conditions for your catfish to thrive. Maintaining water quality through filtration, regular cleaning, and monitoring water parameters is crucial to ensure the health and productivity of your fish. Whether you are farming on a small or large scale, investing in the proper setup and care of tanks will ensure that your catfish farm runs efficiently and successfully.

CHAPTER 4

PROCURING AND STOCKING FINGERLINGS

4.1 Where to Buy Healthy Fingerlings for Your Catfish Farm

The success of your catfish farming venture begins with purchasing healthy fingerlings. Fingerlings are juvenile fish, typically between 1-3 inches in length, that have reached a stage where they can be transferred to grow-out systems such as ponds or tanks. Selecting high-quality fingerlings ensures faster growth rates, lower mortality, and a higher chance of successful farming. In this section, we'll consider where to buy healthy fingerlings, what to look for when purchasing, and the importance of sourcing from reputable suppliers.

1. Understanding the Importance of Healthy Fingerlings

Fingerlings are the foundation of your catfish farm, and the health and quality of these young fish will directly impact your farm's productivity and profitability. Healthy fingerlings grow faster, are more resistant to diseases, and can adapt well to the conditions on your farm. On the other hand, poor-quality or unhealthy fingerlings can lead to high mortality rates, slower growth, and increased costs for medication and management.

2. Sources for Buying Healthy Fingerlings

There are several places where you can purchase fingerlings, each offering different advantages. It's important to choose a reliable source to ensure the fingerlings are disease-free and of the right breed for your farm. Here are the common places to buy fingerlings:

i. Fish Hatcheries: Fish hatcheries are the most common and reliable source for fingerlings. These specialized facilities breed and raise fish in controlled environments, ensuring that the fingerlings are of high quality and free from diseases.

Benefits of Buying from Hatcheries:

- Disease-Free Fingerlings: Hatcheries often have strict biosecurity measures in place to prevent the spread of diseases, ensuring the health of the fish.

- Breeds and Varieties: Hatcheries usually offer a range of catfish breeds to choose from, allowing you to select the one that suits your farm's goals and environment.

- Expert Guidance: Hatcheries are run by professionals who can offer valuable advice on the care and management of the fingerlings, as well as the specific breed you choose.

- Consistent Supply: You can typically order fingerlings in bulk and schedule deliveries based on your farm's needs, ensuring a consistent supply of young fish.

How to Find Hatcheries:

- Local Hatcheries: Look for hatcheries in your region to reduce transport costs and minimize the stress on the fish during travel. Local hatcheries may also offer fish that are better adapted to the climate and water conditions in your area.

- Government Fish Farms: In some regions, government-run fish farms provide fingerlings at subsidized rates to encourage fish farming. These farms often maintain high standards of care and disease prevention.

- Online Hatcheries: Some hatcheries offer fingerlings for sale online, providing nationwide delivery. If using this option, make sure the hatchery is reputable and has proper transport procedures in place to ensure the fingerlings arrive healthy.

ii. Aquaculture Cooperatives and Associations: Aquaculture cooperatives or fish farming associations are groups of farmers or organizations working together to promote fish farming. These cooperatives often have access to healthy fingerlings, which they distribute to their members.

Benefits of Cooperatives:

- Bulk Buying: Cooperatives may offer fingerlings at discounted rates due to bulk purchases, which can save you money.

- Community Support: Being part of a cooperative means you can access shared knowledge, resources, and support from other fish farmers.

iii. University and Research Institutions: Many universities and research institutions have aquaculture programs where they breed and raise fingerlings for research and commercial purposes. These institutions often sell surplus fingerlings to local farmers at reasonable prices.

Advantages of Buying from Research Institutions:

- High-Quality Fingerlings: Since research institutions focus on breeding techniques, genetics, and disease prevention, their fingerlings are often of the highest quality.
- Access to New Varieties: Research programs may develop or improve catfish breeds that have better growth rates, disease resistance, or other desirable traits.
- Expert Advice: Universities and research institutions can offer valuable insights and advice based on their ongoing studies and research in aquaculture.

iv. Private Catfish Farms: Some established catfish farms breed their own fingerlings and sell them to other farmers. These private farms can be a good source, especially if they have a reputation for healthy fish stock.

What to Look For:

- Proven Track Record: Ensure the farm has a good reputation and a history of providing healthy fingerlings.
- Biosecurity Measures: The farm should have strict disease prevention protocols to ensure their fingerlings are not exposed to illnesses.

v. Aquaculture Trade Shows and Expos: Aquaculture trade shows and expos often feature hatcheries, farms, and suppliers showcasing their products, including fingerlings. Attending such events can give you direct access to multiple suppliers, allowing you to compare options and speak with experts in person.

3. Factors to Consider When Buying Fingerlings

When purchasing fingerlings, there are several key factors to keep in mind to ensure you are buying healthy stock:

i. **Breed and Variety:** Choose a breed of catfish that is suitable for your farm's environment and goals. For instance, the African catfish and the Channel catfish are two popular varieties known for their fast growth rates and adaptability to farming conditions. Ensure the hatchery offers the breed you want, and ask about its growth potential, feed requirements, and disease resistance.

ii. **Health and Disease-Free Status:** Healthy fingerlings should be active, free from deformities, and show no signs of disease or infection. When visiting a hatchery or farm, inspect the fingerlings closely for any abnormalities, sluggish behaviour, or external parasites.

- Look for clear eyes, intact fins, and smooth skin.
- Avoid fingerlings that are lethargic or have discoloured patches on their bodies.

iii. **Size and Age:** Fingerlings should be of uniform size and age. This ensures they grow at a similar rate, making it easier to manage them in your farm. Fingerlings that vary significantly in size can lead to bullying, where larger fish dominate smaller ones, leading to uneven growth and stress.

iv. **Transport and Handling:** Ensure that the hatchery or farm follows proper transport and handling procedures. Fingerlings should be packed in oxygenated bags or tanks with adequate water for the journey. Improper handling during transport can lead to stress and high mortality rates.

v. **Documentation:** Ask the supplier for documentation related to the fingerlings, such as health certificates or vaccination records (if applicable). This provides proof that the fish have been raised in a disease-free environment.

4. Building Long-Term Relationships with Suppliers

Building a long-term relationship with a reliable supplier is key to ensuring a steady supply of healthy fingerlings for your farm. A good relationship with a hatchery or supplier will give you access to valuable advice, updates on new breeds, and possibly even discounts or priority service as a repeat customer.

Tips for Building Relationships:

- Communicate Regularly: Keep in touch with your supplier about your farm's needs, challenges, and successes.

- Request Visits: If possible, visit the hatchery or farm to inspect their operations and fish health protocols. This will give you confidence in the quality of their fingerlings.

- Provide Feedback: After purchasing fingerlings, give feedback to the supplier about the health and growth performance of the fish. This helps them improve their services and provides you with better fish in the future.

Buying healthy fingerlings is the first critical step in setting up a successful catfish farm. By sourcing fingerlings from reputable hatcheries, cooperatives, or institutions, and paying attention to health, breed, and transport, you can ensure the long-term productivity and success of your farm. Building strong relationships with suppliers will also help maintain a steady and reliable supply of high-quality fingerlings as your farm grows.

4.2 Transporting Fingerlings Safely

Transporting fingerlings safely is crucial to ensuring the survival and health of these young catfish. Fingerlings are delicate, and improper handling during transport can lead to stress, injury, or even death. By following best practices, you can minimize risks and ensure they arrive at your catfish farm in good condition, ready to grow and thrive.

Why Safe Transport is Important

Fingerlings are in a vulnerable stage of life. They are sensitive to changes in their environment, and the transportation process can be particularly stressful for them. Stress can weaken their immune system, making them more prone to disease. In extreme cases, improper transport can result in high mortality rates.

Safe transportation involves ensuring the fingerlings are kept in optimal conditions throughout the journey, minimizing stress factors such as temperature fluctuations, oxygen depletion, and handling errors. This increases their chances of survival and sets the foundation for healthy growth on your farm.

1. Preparing for Transport

Before transporting fingerlings, it is essential to plan and prepare thoroughly. Preparation involves ensuring the transport containers, water quality, and necessary equipment are ready to provide the best possible environment during transit.

i. Use Proper Transport Containers: The choice of container for transporting fingerlings is crucial. Depending on the number of fish and the distance of transport, you can use:

- Plastic Bags: These are commonly used for shorter trips. The bags are filled with water and oxygen to create a suitable environment for the fish. These bags are usually sealed tightly to prevent leaks and oxygen loss.

- Transport Tanks: For longer distances or larger quantities of fingerlings, portable tanks or containers with aeration systems are used. These tanks ensure proper water quality and oxygen levels throughout the journey.

- Insulated Containers: Insulated transport tanks or containers help maintain stable water temperatures, preventing sudden changes that can stress the fingerlings.

ii. Prepare Clean, Oxygenated Water: The water used for transporting fingerlings should be clean and free of contaminants. The key factors to consider include:

- Oxygen Levels: High oxygen levels are crucial during transport, as fingerlings use oxygen quickly, especially when confined in a small volume of water. If using plastic bags, they should be filled with oxygen (typically around 1/3 water and 2/3 oxygen). For tanks, use aeration systems or oxygenation equipment to maintain adequate levels.

- Water Quality: The water in the transport containers should be free of pollutants and toxins. Ideally, it should be sourced from a known, clean environment or the same source the fingerlings were raised in to minimize stress from water changes.

iii. Temperature Control: Temperature fluctuations can shock the fingerlings, so it's important to maintain a consistent water temperature during transport.

- Cooler temperatures: For shorter transport times, slightly cooler water (a few degrees lower than the pond or tank temperature) can reduce the metabolic rate of the fish, helping them conserve oxygen and reduce stress.

- Avoid Overheating: If transporting in warm conditions, ensure the transport container is shaded or insulated to prevent overheating.

2. Transporting Fingerlings Safely

Once your fingerlings are in the transport containers, keeping the environment stable and stress-free is crucial. Follow these steps for a safe journey:

i. Avoid Overcrowding: Overcrowding fingerlings during transport can lead to oxygen depletion, increased waste production, and stress. The transport container should have enough space for the fish to move around without being compressed against each other.

As a general rule:

- For short trips (up to 2-3 hours), use approximately 100-200 fingerlings per 5 gallons of water.

- For longer trips, reduce the number to ensure better oxygen levels and water quality.

ii. Monitor Oxygen Levels: Fingerlings consume oxygen quickly, especially when stressed during transport. If you are using plastic bags, ensure that they are filled with pure oxygen rather than regular air. For transport tanks, use an air pump or oxygen diffuser to keep oxygen levels high throughout the journey.

- Check the oxygen system: Before setting out, test the oxygenation equipment to make sure it's functioning properly. If using bags, periodically check the fish for signs of oxygen depletion (such as gasping or swimming at the surface).

iii. Minimize Handling: Fingerlings are fragile and should be handled as little as possible to avoid injuries or added stress. When moving them from their holding tanks to transport containers, use soft, fine-mesh nets to gently transfer the fish. Avoid touching the fish with your hands, as this can remove their protective mucus layer, leaving them vulnerable to infection.

iv. Stable Transport: During transport, ensure that the containers remain stable and are not subjected to sudden jolts or movements. This helps keep the fingerlings calm and reduces the risk of injury. Secure the containers properly in your vehicle, and drive carefully to avoid rough handling.

v. Monitor During Transit: If the trip is long, stop periodically to check on the fingerlings. Ensure the oxygen levels are adequate and that the water temperature has not risen or dropped significantly. Adjustments can be made en route to ensure the fish remain in good condition.

3. Acclimatizing Fingerlings to Your Farm

Once the fingerlings have safely arrived at your farm, the next critical step is acclimatizing them to their new environment. Sudden changes in water temperature, pH, or quality can shock the fish and lead to fatalities. Proper acclimatization helps the fingerlings adjust smoothly, reducing the risk of stress or disease.

i. Gradual Acclimatization: The process of acclimatization should be done gradually, allowing the fingerlings to slowly get used to the water in your pond or tank. Follow these steps:

- Temperature Matching: If the transport water temperature differs from the pond or tank water, float the closed transport bag in the pond for 15-30 minutes to allow the water temperatures to equalize.

- Mix Water Slowly: After temperature adjustment, slowly add small amounts of pond water into the transport bag over a period of 15-20 minutes. This helps the fingerlings adjust to the pH, salinity, and other water parameters.

- Release the Fingerlings: Once acclimatized, gently release the fingerlings into the pond or tank. Avoid pouring them out abruptly, as this can cause further stress. Allow them to swim out on their own by slowly tipping the container or bag.

ii. Monitor Post-Release Behaviour: After releasing the fingerlings, observe them closely for the first few hours. Healthy fingerlings will swim actively and explore their new environment. Watch for signs of stress or distress, such as floating, gasping at the surface, or erratic swimming. If any problems arise, it may indicate an issue with the water quality or acclimatization process.

4. Transporting Fingerlings in Extreme Weather

In extreme weather conditions, such as very hot or cold temperatures, extra precautions are necessary to protect the fingerlings. Here are a few tips:

- Hot Weather: If transporting in high temperatures, keep the transport containers in a shaded area and avoid direct sunlight. Insulated coolers or tanks can help keep the water temperature stable. You can also use ice packs placed on the outside of the container (not inside the water) to maintain a cooler temperature.

- Cold Weather: During cold weather, use insulated containers to prevent the water from becoming too cold. You can also use portable water heaters or warm blankets around the containers to maintain a suitable temperature.

5. Building a Transport Plan for Future Shipments

As your catfish farm grows, you may need to transport fingerlings multiple times, either to restock or expand your operations. Having a well-structured transport plan will ensure the safe and efficient delivery of fingerlings over time.

Key points to include in your plan:

- Reliable Transport Equipment: Invest in high-quality transport containers, oxygen systems, and insulation materials that are reusable and durable.

- Supplier and Farm Communication: Maintain clear communication with the hatchery or supplier to ensure fingerlings are ready for transport at the right time and in the best condition.

- Scheduled Deliveries: Plan deliveries during the coolest parts of the day (early morning or late afternoon) to avoid extreme temperatures.

Transporting fingerlings safely is an essential part of starting and maintaining a successful catfish farm. By using the right equipment, maintaining stable water quality, and handling the fingerlings carefully, you can reduce stress and increase their chances of survival. Proper acclimatization once they arrive at your farm ensures that they adapt smoothly to their new environment, setting them up for healthy growth and a productive farming operation. With the right approach, your fingerlings will have the best possible start on your farm.

4.3 Stocking Your Pond or Tanks

Stocking your pond or tanks with fingerlings is a key step in starting a catfish farm. This process involves introducing the right number of young fish into your pond or tanks, providing them with a suitable environment for growth, and ensuring their long-term health. Proper stocking ensures that your catfish have enough space, oxygen, and resources to thrive, reducing competition and increasing survival rates.

1. Understanding the Stocking Process

Stocking refers to the act of introducing fingerlings into a new environment, whether that's a pond, tank, or other water system. This step is crucial for the success of your catfish farm as it directly influences the growth rate, health, and overall production of your farm.

To do this successfully, you need to consider several factors, such as the number of fish to stock, the size of your pond or tank, and the quality of the environment you've prepared for them.

2. Determining the Stocking Density

Stocking density refers to the number of fingerlings you introduce per unit of water. It is one of the most important factors when stocking your pond or tanks, as overcrowding can lead to oxygen depletion, increased waste production, and a higher risk of disease. On the other hand, understocking can result in wasted resources and inefficient use of space.

i. Ideal Stocking Density: The ideal stocking density depends on whether you're using a pond or tank system, the size of your water body, and your farming goals. A common guideline for catfish farming is:

- For ponds: Stock around 1,500 to 2,500 fingerlings per acre of pond surface, depending on your aeration system and water quality.
- For tanks or recirculating systems: The general recommendation is 50-100 fingerlings per cubic meter of water, with a good aeration and filtration system in place.

When deciding on stocking density, keep the following factors in mind:

- Water Quality: The better your water quality (especially oxygen levels), the higher your stocking density can be.
- Growth Goals: Higher densities can produce more catfish, but they may grow more slowly compared to those in lower-density environments.

ii. Overcrowding vs. Understocking

- Overcrowding: When too many fingerlings are stocked, it can lead to overcrowding. Overcrowding increases stress, oxygen demand, and waste production, which can negatively affect the fish's health. The risk of diseases spreading quickly is also higher in densely stocked environments.
- Understocking: Stocking too few fingerlings might seem like a safer approach, but it can result in underutilizing the available space and resources in your pond or tank. Understocking can lead to slow production rates and lower profitability in the long run.

3. Timing of Stocking

The timing of when you introduce fingerlings into your pond or tanks is another key consideration. Stocking should be done when environmental conditions are optimal for the fingerlings to adapt and grow.

i. Best Time of Year: For outdoor ponds, it's best to stock catfish fingerlings in the spring or early summer when water temperatures are moderate and stable. Catfish thrive in warm water, ideally between 75°F and 85°F (24°C - 29°C). Stocking during this period allows the fingerlings to grow rapidly during the warmer months.

For indoor tanks or recirculating systems, the timing is less critical since water temperatures can be controlled. However, it's still important to ensure that the water conditions are stable and suitable before introducing the fish.

4. Preparing the Pond or Tank

Before introducing your fingerlings, it's essential to ensure that the pond or tank environment is ready for them. This includes making sure the water quality is optimal, and the environment is free of harmful substances or predators.

i. Water Quality Testing: Before stocking, check the water quality to ensure it meets the following requirements:

- pH Levels: Maintain a pH level between 6.5 and 8.5, which is ideal for catfish.

- Dissolved Oxygen: Ensure sufficient oxygen levels, typically 4-5 mg/L, especially for ponds or tanks with higher stocking densities. If needed, install aeration systems to maintain adequate oxygen levels.

ii. Predator Control: If you're stocking a pond, it's crucial to ensure that predators such as birds, snakes, or other larger fish species don't have access to your catfish fingerlings. Consider using predator nets or fencing to protect the young fish.

iii. Water Conditioning: For newly constructed ponds or tanks, it may take some time for the water to establish a balanced ecosystem. Allow time for the water to "settle" and develop the right balance of microorganisms before stocking. Additionally, if you've added new water to a pond, give it time to adjust to the right temperature and pH before introducing the fingerlings.

5. Acclimatizing Fingerlings Before Stocking

Just as with transporting fingerlings, acclimatizing them to the new environment is crucial before releasing them into the pond or tanks. Sudden changes in water temperature, pH, or oxygen levels can shock the fingerlings, leading to stress or death.

i. Gradual Acclimatization

To acclimatize fingerlings to their new environment:

- Float the Bag: If your fingerlings arrive in plastic bags, float the sealed bags in the pond or tank for 15-30 minutes to allow the water temperatures to equalize.

- Mix Water Gradually: Slowly introduce small amounts of pond or tank water into the bag over the next 10-15 minutes. This allows the fingerlings to adjust to changes in pH, temperature, and water chemistry gradually.

- Release the Fingerlings: Once the temperature and water conditions have equalized, gently release the fingerlings into the pond or tank by slowly tipping the container. Avoid pouring them out forcefully to reduce stress.

6. Post-Stocking Care

After stocking your pond or tank, monitor the fingerlings closely for the first few days to ensure they are adjusting well. Here's what to look out for:

i. Monitor Behaviour: Healthy fingerlings should swim actively and explore their new environment. Watch for any signs of stress or illness, such as floating, gasping at the surface, or erratic swimming. If any issues arise, they may indicate problems with the water quality or the acclimatization process.

ii. Maintain Water Quality: Regularly test the water quality, especially during the first few weeks after stocking. Keep an eye on dissolved oxygen levels, pH, and temperature to ensure the environment remains stable.

iii. Initial Feeding: Feed the fingerlings within 24-48 hours of stocking. Use high-quality, nutrient-rich feed designed for young catfish. Start with small quantities and gradually increase the feed as the fish settle and grow. Avoid overfeeding, as uneaten food can decay and pollute the water.

Stocking your pond or tanks with fingerlings is a critical step that sets the stage for the success of your catfish farm. By following best practices, such as determining the right stocking density, acclimatizing the fingerlings properly, and maintaining optimal water conditions, you can ensure that your catfish have the best possible start. With proper care and attention, your stocked fingerlings will grow into healthy, mature catfish, leading to a thriving and productive farm.

CHAPTER 5

FEEDING AND NUTRITION

5.1 Understanding Catfish Dietary Needs

Feeding your catfish the right diet is one of the most important aspects of running a successful catfish farm. Catfish have specific nutritional requirements that must be met to ensure they grow quickly, stay healthy, and develop into high-quality fish for market or consumption. Understanding their dietary needs helps farmers choose the right feed, establish feeding schedules, and ultimately improve the efficiency and profitability of their farm.

1. The Basics of Catfish Nutrition

Like all animals, catfish require a balanced diet that provides them with essential nutrients, including proteins, carbohydrates, fats, vitamins, and minerals. Each of these nutrients plays a vital role in the fish's growth, immune system, and overall health.

i. Proteins

- Role: Proteins are the building blocks of growth. They are essential for muscle development, tissue repair, and other bodily functions.

- Sources: In catfish feed, protein typically comes from both animal and plant sources, such as fish meal, soybean meal, and corn gluten.

- Requirement: Young catfish (fingerlings) need a high-protein diet (around 35-40%) to support rapid growth, while mature catfish need slightly less (about 28-32%).

ii. Carbohydrates

- Role: Carbohydrates provide energy that catfish use for swimming, digestion, and other daily activities.
- Sources: Common sources of carbohydrates in catfish feed include grains like corn, wheat, and rice bran.
- Requirement: Catfish can efficiently use carbohydrates for energy, but their diet should be balanced so that protein isn't wasted for energy production. A moderate amount of carbohydrates (20-30%) is ideal.

iii. Fats (Lipids)

- Role: Fats are a concentrated energy source and are necessary for the absorption of fat-soluble vitamins (A, D, E, K). They also help with maintaining healthy skin and scales.
- Sources: Fish oils, soybean oil, and poultry fat are commonly used in catfish feeds.
- Requirement: Catfish need about 5-10% fat in their diet. Too little fat can result in poor growth, while too much can lead to fatty deposits in the fish.

iv. Vitamins and Minerals

- Role: Vitamins and minerals are essential for the overall well-being of catfish. They support functions like bone development, metabolic processes, and disease resistance.
- Sources: Most commercial catfish feeds are fortified with essential vitamins (such as vitamins A, D, E, and C) and minerals like calcium and phosphorus.
- Requirement: While vitamins and minerals are required in small amounts, they are critical to preventing deficiencies that could lead to poor growth, weakened immune systems, and skeletal deformities.

2. Types of Feed for Catfish

There are various types of feed available for catfish, each designed to meet their nutritional needs at different stages of life.

The choice of feed is influenced by the age, size, and growth goals of your fish.

i. Commercial Pelleted Feed

- Description: The most common type of feed used in catfish farming is pelleted feed, which comes in floating and sinking varieties. These pellets are specially formulated to provide all the necessary nutrients in balanced proportions.

- Floating Pellets: These float on the water's surface, making it easier to monitor how much the fish are eating. They are ideal for fingerlings and smaller catfish, which feed at the surface.

- Sinking Pellets: These sink to the bottom of the pond or tank, where larger catfish tend to feed. Sinking pellets are best used for mature or bottom-dwelling catfish.

ii. Supplemental Feeds

- Description: In some cases, farmers may choose to supplement commercial feed with natural food sources or cheaper alternatives, especially in extensive farming systems.

- Natural Sources: Catfish are omnivorous and can feed on insects, small fish, algae, and other aquatic organisms found in ponds. These can act as a supplementary food source.

- Agricultural By-products: Some farms use by-products like rice bran, maize bran, or kitchen scraps to reduce feed costs. While these may provide some nutrition, they are typically not sufficient as a sole diet.

iii. Feed Additives

- Description: Additives like probiotics, enzymes, or fish oils may be added to commercial feeds to enhance growth, improve digestion, or boost immunity. They are not always necessary but can be beneficial in intensive farming setups.

3. Feeding Habits of Catfish

Catfish have different feeding habits depending on their species, size, and environment. Understanding these habits can help you plan effective feeding strategies that promote healthy growth and minimize waste.

i. Feeding Behaviour

- Young Catfish (Fingerlings): Fingerlings are surface feeders, which is why floating pellets are ideal for them. They eat more frequently as they are rapidly growing and require a constant supply of nutrients.
- Adult Catfish: As catfish mature, their feeding behaviour shifts to the bottom of the pond or tank, especially for larger species. They are more likely to consume sinking pellets and tend to feed less frequently than juveniles.

ii. Feeding Frequency

The frequency of feeding depends on the age of the fish and environmental factors like water temperature:

- Fingerlings: These need to be fed at least 2-3 times per day because of their rapid growth.
- Adult Catfish: For mature catfish, feeding once or twice a day is generally sufficient. In colder temperatures, their metabolism slows, and they may need to be fed less frequently.

iii. Feed Conversion Ratio (FCR)

FCR is a measure of how efficiently catfish convert feed into body mass. A low FCR means the fish are using the feed efficiently and growing well, which helps reduce feed costs. Most catfish have an FCR of around 1.5 to 2, meaning they need 1.5 to 2 pounds of feed to gain 1 pound of body weight.

4. Adjusting Feed Based on Growth Stages

Catfish undergo different growth stages, and their dietary needs change as they develop. Feeding them appropriately at each stage helps optimize their growth.

i. Fingerlings (Young Fish)

- Nutritional Needs: Fingerlings require a high-protein diet (35-40%) to support their rapid growth and development. High-quality commercial feed designed specifically for young fish is recommended.

- Feeding Strategy: Feed them small amounts multiple times per day to keep them well-nourished and prevent wastage.

ii. Grow-Out Stage

- Nutritional Needs: During the grow-out stage, when the fish are maturing, they require slightly less protein (30-35%) and a balanced mix of carbohydrates and fats. The focus is on maintaining steady growth until they reach market size.

- Feeding Strategy: Depending on the size of your pond or tank and the stocking density, feed them once or twice daily with pelleted feed that meets their energy and protein needs.

iii. Market-Size or Mature Catfish

- Nutritional Needs: Fully grown catfish require less protein (28-32%) but still need a well-balanced diet to maintain their health and weight. At this stage, they are primarily focusing on maintaining their weight rather than growing.

- Feeding Strategy: Feeding once a day is usually sufficient for mature fish. Sinking pellets are often used since these fish tend to feed closer to the bottom of the pond or tank.

5. Monitoring and Managing Feeding Practices

Proper feeding management is essential to ensure that your catfish are healthy and that you're not wasting feed. Overfeeding or underfeeding can have negative effects on your fish and the overall health of your pond or tank.

i. Avoid Overfeeding: Overfeeding can lead to uneaten feed accumulating at the bottom of the pond or tank, which can decompose and pollute the water. This can result in poor water quality, increased ammonia levels, and even diseases. Feed only what the catfish will consume in a short time (usually 10-15 minutes).

ii. Observing Feeding Patterns: Pay attention to the feeding behaviour of your catfish. If they stop eating or show less interest in feed, it could be a sign of illness, stress, or poor water conditions. Regularly observing how your catfish eat helps you adjust feeding amounts and detect any potential problems early.

6. Water Temperature and Feeding

Catfish are ectothermic, meaning their metabolism is directly affected by water temperature. In warmer months, catfish are more active and will eat more. However, as water temperatures drop (below 70°F or 21°C), their metabolism slows, and they require less food. In colder conditions, it may even be necessary to stop feeding until the water temperature rises again.

Understanding the dietary needs of catfish is fundamental to the success of your farm. By providing the right nutrients in the correct proportions, monitoring feeding behaviour, and adjusting the feeding strategy as your fish grow, you can ensure that your catfish thrive. With proper feeding management, you'll not only enhance the growth and health of your fish but also optimize the efficiency and profitability of your catfish farming operation.

5.2 Feeding Schedules and Techniques

Feeding catfish properly is crucial for their growth, health, and overall productivity on your farm. Establishing effective feeding schedules and techniques ensures that your fish receive the right amount of nutrients at the right times, which helps optimize their growth and reduce waste. This section will guide you through the best practices for feeding schedules and techniques to keep your catfish farm running smoothly.

1. Developing a Feeding Schedule

A well-planned feeding schedule helps manage feed costs and maintain the health of your catfish. The schedule you establish will depend on the age, size, and environmental conditions of your fish.

i. Feeding Frequency

The frequency of feeding varies with the age and size of the catfish, as well as environmental factors such as water temperature.

- Fingerlings (Young Fish):
    - ✓ Frequency: Feed 2-3 times per day. Young catfish have high metabolic rates and need frequent feedings to support their rapid growth.
    - ✓ Why: Multiple feedings throughout the day ensure they get enough nutrients and reduce competition for food.
- Grow-Out Stage:
    - ✓ Frequency: Feed 1-2 times per day.
    - ✓ Why: As catfish grow, their growth rate slows down, and they require fewer feedings. This helps in maintaining steady growth while preventing overfeeding.
- Market-Size or Mature Catfish:
    - ✓ Frequency: Feed once or twice per day.
    - ✓ Why: Mature catfish have slower growth rates and less frequent feeding helps maintain their weight without wasting feed.

ii. Adjusting for Water Temperature

Water temperature significantly affects catfish metabolism and feeding behaviour.

- Warm Water (70°F or 21°C and above):
    - ✓ Feeding: Catfish are more active and will eat more. Maintain the regular feeding schedule as their appetite increases.
- Cold Water (below 70°F or 21°C):
    - ✓ Feeding: Catfish's metabolism slows down, and their feeding rate decreases. Reduce the amount of feed and frequency, and monitor their feeding

closely. In very cold temperatures, it may be necessary to reduce or stop feeding.

2. Effective Feeding Techniques

The method used to deliver feed can affect how well your catfish utilize it and how much waste is produced.

i. Manual Feeding

- Description: Feed is distributed by hand or with simple feeding devices.
- When to Use: Ideal for small-scale operations or when you need precise control over feed distribution.
- Benefits: Allows you to monitor the amount of feed used and observe fish behaviour closely.
- Limitations: Can be labour-intensive and less practical for large farms.

ii. Automatic Feeders

- Description: Devices that automatically dispense feed at scheduled times and amounts.
- When to Use: Suitable for medium to large-scale operations where regular, consistent feeding is required.
- Benefits: Reduces labour, ensures consistent feeding times, and can be programmed to adjust feed amounts based on fish size and growth stages.
- Limitations: Initial cost can be high, and maintenance is required to prevent feed blockages and ensure proper function.

iii. Broadcast Feeding

- Description: Feed is spread over a wide area, either manually or using machinery.
- When to Use: Useful for ponds with large surface areas where fish are spread out.
- Benefits: Ensures that feed reaches all fish, particularly in extensive farming systems.

- Limitations: Can lead to uneven feed distribution and potential waste if not managed properly.

iv. Spot Feeding

- Description: Feed is placed in specific areas of the pond or tank.
- When to Use: Ideal for tanks or ponds with high-density fish populations.
- Benefits: Reduces waste and ensures that fish in the targeted area receive the feed.
- Limitations: Requires careful monitoring to ensure all fish get adequate feed.

3. Monitoring and Adjusting Feeding Practices

Regular monitoring helps ensure that your feeding practices are effective and that your catfish are receiving the appropriate amount of nutrition.

i. Observing Fish Behaviour

- Feeding Response: Watch how quickly and eagerly the fish consume the feed. If they are not eating as expected, it could indicate problems such as poor water quality or illness.
- Adjustments: Based on their feeding response, you may need to adjust feed amounts, change feeding frequency, or switch feed types.

ii. Monitoring Feed Waste

- Feed Waste: Uneaten feed can accumulate at the bottom of the pond or tank, leading to water quality issues. Adjust feeding amounts to minimize waste.
- Water Quality: Regularly check water quality parameters like ammonia, nitrite, and oxygen levels, as poor water quality can affect feeding efficiency and fish health.

iii. Record Keeping

- Feeding Records: Keep detailed records of feeding times, amounts, and any changes in feeding practices. This helps track the growth and health of your fish and makes it easier to adjust practices as needed.

4. Troubleshooting Common Feeding Issues

Even with the best practices, you may encounter issues related to feeding. Here are some common problems and solutions:

i. Fish Not Eating

- Possible Causes: Poor water quality, stress, disease, or inappropriate feed.
- Solutions: Check water quality and make necessary adjustments. Ensure that the feed is suitable for the fish's age and size. If disease is suspected, consult a veterinarian.

ii. Excessive Feed Waste

- Possible Causes: Overfeeding or feed not suitable for the fish's appetite.
- Solutions: Adjust feeding amounts and ensure that the feed is appropriate for the fish. Implement feeding techniques that minimize waste, such as spot feeding.

iii. Uneven Growth

- Possible Causes: Inconsistent feeding or competition among fish.
- Solutions: Ensure even distribution of feed and monitor for aggressive behaviour among fish. Adjust feeding techniques and amounts as needed to support balanced growth.

Establishing effective feeding schedules and techniques is crucial for the success of your catfish farm. By tailoring your feeding practices to the age and size of your fish, adjusting for environmental factors, and employing the right feeding methods, you can promote optimal growth, reduce waste, and ensure the health and productivity of your catfish. Regular monitoring and adjustments based on observed behaviour and feeding patterns will help you maintain a successful and efficient catfish farming operation.

5.3 Managing Feed Costs and Efficiency

Effective management of feed costs and efficiency is crucial for the financial success of your catfish farm. Feed typically represents a significant portion of your operational expenses, so optimizing feed use and controlling costs can have a major impact on your profitability. This section provides a detailed guide on strategies to manage feed costs and improve feeding efficiency.

1. Understanding Feed Costs

Feed costs are a major expense in catfish farming and can vary based on feed type, quality, and market prices. To manage these costs effectively, it's important to understand the factors influencing feed prices and how to make informed choices.

i. Types of Feed

- Commercial Feeds: These are pre-formulated feeds that come in various types, such as pellets, crumbles, and powders. They are designed to meet the nutritional needs of catfish at different growth stages.
    - ✓ Advantages: Balanced nutrition, convenience, and often include additives for health and growth.
    - ✓ Disadvantages: Higher cost compared to homemade or alternative feeds.
- Homemade Feeds: These are feeds prepared on the farm using local ingredients. They can be customized to meet specific nutritional needs.
    - ✓ Advantages: Potentially lower cost and tailored nutrition.
    - ✓ Disadvantages: Requires careful formulation and quality control to ensure nutritional adequacy.
- Alternative Feeds: Includes agricultural by-products, fish meal, and other non-traditional feed sources.

- ✓ Advantages: May be more cost-effective and locally available.
- ✓ Disadvantages: Nutritional quality can vary, and they may require additional supplementation.

ii. Factors Affecting Feed Prices

- Ingredient Costs: Prices of ingredients like fishmeal, soybeans, and corn can fluctuate based on market conditions.
- Production Costs: The cost of manufacturing feed, including labour, energy, and processing, can impact feed prices.
- Transportation Costs: Costs associated with transporting feed to your farm can affect overall feed expenses.

2. Strategies for Reducing Feed Costs

Effective strategies can help you manage and reduce feed costs while maintaining the health and productivity of your catfish.

i. Optimizing Feed Formulation

- Custom Formulation: Work with a nutritionist or feed specialist to develop a feed formulation that meets the nutritional needs of your catfish while minimizing costs.
- Ingredient Substitution: Use cost-effective, locally available ingredients that still provide essential nutrients.

ii. Improving Feed Conversion Ratio (FCR)

- Definition: FCR is a measure of how efficiently fish convert feed into body weight. A lower FCR indicates better feed efficiency.
- Monitoring: Regularly monitor FCR to assess the effectiveness of your feeding practices. Aim for an optimal FCR to reduce feed waste and costs.

- Optimization: Adjust feed types, feeding schedules, and environmental conditions to improve FCR. Ensuring that the feed is of high quality and appropriate for the fish's growth stage can also help.

iii. Implementing Feeding Management Techniques

- Feeding Rates: Avoid overfeeding by adjusting feed amounts based on fish size, water temperature, and growth stage. Overfeeding leads to waste and increased costs.
- Feeding Methods: Use feeding techniques such as spot feeding or automatic feeders to reduce feed wastage and ensure that all fish receive adequate nutrition.

3. Enhancing Feed Efficiency

Improving feed efficiency helps to make the most out of the feed you purchase, which can lower overall costs and boost productivity.

i. Monitoring Fish Growth

- Regular Checks: Measure and record fish growth regularly to assess the effectiveness of your feeding practices. Growth rates can help you determine if adjustments are needed in feed amounts or types.
- Adjustments: Make necessary adjustments based on growth data to optimize feed use and ensure that fish are growing at the desired rate.

ii. Improving Water Quality

- Water Quality: Maintain optimal water quality to support healthy fish growth and efficient feed utilization. Poor water quality can affect digestion and feed conversion.
- Monitoring: Regularly test water parameters such as pH, oxygen levels, and ammonia concentrations. Make adjustments to maintain ideal conditions.

iii. Regular Maintenance

- Feed Storage: Store feed properly to prevent spoilage and loss of nutritional value. Use airtight containers and keep feed in a dry, cool place.
- Equipment: Regularly maintain feeding equipment to ensure consistent and efficient feed distribution.

4. Evaluating and Adjusting Feeding Practices

Continually evaluating your feeding practices helps to identify areas for improvement and make necessary adjustments to manage feed costs effectively.

i. Analysing Feed Records

- Tracking: Keep detailed records of feed purchases, usage, and costs. Analysing these records can help you identify trends and areas for cost savings.
- Reviewing: Regularly review feed records to assess if adjustments in feed types, amounts, or suppliers are needed.

ii. Consulting Experts

- Nutritionists: Work with feed nutritionists to refine feed formulations and improve feed efficiency.
- Farm Advisors: Consult with agricultural or aquaculture advisors for insights into cost-saving measures and best practices.

Managing feed costs and efficiency is a critical component of running a successful catfish farm. By understanding feed types and costs, optimizing feed formulation and practices, and continuously monitoring and adjusting your feeding strategies, you can reduce expenses and improve the overall productivity of your farm. Implementing effective feed management techniques and leveraging expert advice will help you achieve long-term success and profitability in catfish farming.

CHAPTER 6

MANAGING WATER QUALITY

6.1 Importance of Water Quality in Catfish Farming

Water quality is a fundamental aspect of successful catfish farming. It directly affects the health, growth, and productivity of your fish, as well as the overall efficiency of your farm. Maintaining high water quality is essential for ensuring that your catfish thrive and your farming operation remains sustainable and profitable. Here's a comprehensive overview of why water quality is so crucial in catfish farming.

1. Impact on Fish Health

i. Disease Prevention

- Clean Water: Good water quality helps prevent the outbreak of diseases and parasites. Polluted or poor-quality water can stress catfish, making them more susceptible to infections and diseases.

- Pathogen Control: High levels of ammonia, nitrites, and other harmful substances can create an environment where pathogens thrive. Maintaining proper water quality helps control pathogen levels and reduces disease risk.

ii. Stress Reduction

- Comfortable Environment: Catfish are sensitive to changes in their environment. Poor water quality can lead to stress, which weakens their immune system and can lead to health problems. Ensuring optimal water conditions helps keep fish stress-free and healthy.

2. Growth and Development

i. Optimal Growth Rates

- Nutrient Absorption: High-quality water ensures that catfish can efficiently absorb nutrients from their feed. Poor water quality can affect digestion and nutrient uptake, leading to slower growth and lower feed conversion rates.
- Healthy Development: Consistent water quality supports normal growth and development. This includes maintaining appropriate levels of oxygen, temperature, and pH, which are critical for healthy fish growth.

ii. Feed Efficiency

- Reduced Waste: Maintaining clean water helps to minimize feed waste. High levels of organic matter from decomposing feed can degrade water quality and affect feed efficiency. Good water quality helps ensure that more of the feed is utilized by the fish rather than wasted.

3. Environmental Stability

i. Water Parameters

- pH Levels: The pH of the water affects fish health and metabolic processes. Catfish generally thrive in slightly acidic to neutral water (pH 6.5-8.0). Extreme pH levels can stress the fish and affect their growth.
- Temperature: Water temperature influences metabolic rates and overall fish health. Catfish have specific temperature ranges for optimal growth, and fluctuations outside this range can impact their well-being.

ii. Oxygen Levels

- Dissolved Oxygen: Adequate levels of dissolved oxygen are crucial for fish respiration and overall health. Low oxygen levels can lead to hypoxia, which can be harmful or even fatal to fish. Proper aeration and water circulation help maintain sufficient oxygen levels.

4. Water Quality Management

i. Regular Monitoring

- Testing: Regularly test water parameters such as pH, ammonia, nitrite, nitrate, dissolved oxygen, and temperature. Monitoring these parameters helps identify potential issues before they become serious problems.

- Record Keeping: Keep detailed records of water quality tests and maintenance activities. This helps track changes over time and assists in identifying patterns or trends.

ii. Filtration and Aeration

- Filtration Systems: Use appropriate filtration systems to remove debris, excess nutrients, and waste from the water. Effective filtration helps maintain clean water and reduces the risk of pollution.

- Aeration Devices: Install aeration devices to ensure adequate oxygen levels in the water. Aerators help maintain oxygen levels and improve water circulation.

iii. Water Exchange and Treatment

- Water Exchange: Regularly replace a portion of the water to dilute pollutants and maintain water quality. The frequency and amount of water exchange depend on the size of your system and the load of fish.

- Water Treatment: Use water treatments as needed to address specific issues such as high ammonia levels or low pH. Ensure that any treatments used are safe for catfish and follow recommended guidelines.

5. Economic Considerations

i. Cost of Water Management

- Investment: Investing in water quality management systems, such as filtration and aeration, can be costly but is essential for the health and productivity of your catfish. Consider these costs as part of your overall farm budget.

- Cost Savings: Proper water management can lead to long-term cost savings by reducing disease outbreaks, improving feed efficiency, and enhancing growth rates. Investing in water quality can ultimately improve the profitability of your farm.

ii. Risk Mitigation

- Minimizing Losses: Poor water quality can lead to significant losses in fish health and production. Effective water management reduces the risk of such losses and helps ensure the sustainability of your farming operation.

Water quality is a cornerstone of successful catfish farming. It affects every aspect of fish health, growth, and productivity. By understanding the importance of water quality and implementing effective management practices, you can create an optimal environment for your catfish, minimize health risks, and enhance the overall efficiency and profitability of your farm. Regular monitoring, appropriate filtration and aeration, and proactive water management are key to maintaining high water quality and ensuring the success of your catfish farming operation.

6.2 Regular Water Testing

Regular water testing is essential for maintaining optimal water quality in catfish farming. It helps ensure that the aquatic environment remains healthy for your fish, supports their growth, and minimizes the risk of disease. Here's a comprehensive guide to understanding why regular water testing is crucial and how to effectively implement it in your catfish farming operation.

1. Why Regular Water Testing is Important

i. Health and Well-being of Catfish

- Detecting Issues Early: Regular testing helps identify potential problems with water quality before they affect fish health. This early detection allows you to take corrective actions to prevent stress and disease in your catfish.
- Maintaining Optimal Conditions: Different water parameters, such as pH, temperature, and oxygen levels, have specific ranges that are ideal for catfish. Regular testing ensures that these conditions remain within the optimal range, supporting the health and growth of your fish.

ii. Preventing Disease Outbreaks

- Managing Contaminants: Poor water quality can lead to the accumulation of harmful substances like ammonia and nitrites, which can stress fish and promote the growth of pathogens. Regular testing helps manage these contaminants and reduces the risk of disease outbreaks.
- Controlling Algae Growth: Testing for nutrient levels can help prevent excessive algae growth, which can lead to oxygen depletion and other water quality issues.

iii. Improving Feed Efficiency

- Optimizing Growth: High water quality supports efficient feed utilization, which means that fish can grow faster and more effectively. Regular testing helps maintain the right conditions for optimal feed conversion and growth rates.

2. Parameters to Test

To effectively monitor water quality, focus on testing the following key parameters:

i. pH Levels

- Importance: The pH level measures the acidity or alkalinity of the water. Catfish typically thrive in slightly acidic to neutral water (pH 6.5-8.0). Extreme pH levels can stress fish and affect their health.

- Testing Frequency: Test pH levels regularly, at least once a week or more frequently if you notice changes in fish behaviour or water conditions.

ii. Temperature

- Importance: Water temperature affects fish metabolism, growth, and overall health. Catfish have specific temperature ranges for optimal growth, typically between 75-85°F (24-29°C).
- Testing Frequency: Monitor water temperature daily, especially if there are significant fluctuations or changes in the weather.

iii. Dissolved Oxygen (DO)

- Importance: Dissolved oxygen is crucial for fish respiration. Low oxygen levels can lead to hypoxia, which can be harmful or fatal to fish.
- Testing Frequency: Test DO levels regularly, at least once a week, and more frequently if you observe signs of oxygen depletion, such as fish gasping at the surface.

iv. Ammonia, Nitrites, and Nitrates

- Importance: These compounds are products of fish waste and decomposing organic matter. High levels of ammonia and nitrites can be toxic to fish, while nitrates can accumulate and affect water quality.
- Testing Frequency: Test ammonia and nitrites weekly, and nitrates monthly or as needed. Regular testing helps ensure that these levels remain within safe ranges.

v. Alkalinity and Hardness

- Importance: Alkalinity helps buffer pH fluctuations, while hardness affects the availability of calcium and other minerals. Both factors influence the stability of water conditions.

- Testing Frequency: Test alkalinity and hardness monthly or as needed to maintain stable water conditions.

3. How to Perform Water Testing

i. Choosing the Right Testing Equipment

- Test Kits: Use reliable test kits or meters for measuring pH, dissolved oxygen, ammonia, nitrites, and other parameters. Choose kits that are specifically designed for aquaculture and follow the manufacturer's instructions for accurate results.
- Digital Meters: For more precise measurements, consider investing in digital meters for parameters like pH, dissolved oxygen, and temperature.

ii. Sampling Techniques

- Proper Sampling: Collect water samples from different locations and depths to get an accurate representation of the overall water quality. Avoid collecting samples from areas with visible debris or contamination.
- Sample Handling: Use clean, contaminant-free containers for water samples. If testing is not immediate, store samples according to the instructions provided with your test kits.

iii. Interpreting Results

- Understanding Readings: Compare test results with the optimal ranges for catfish. Use reference guides or consult with aquaculture experts if you're unsure about the implications of your results.
- Taking Action: Based on the test results, take appropriate actions to correct any water quality issues. This may include adjusting pH, improving aeration, or performing water changes.

4. Developing a Water Testing Schedule

i. Regular Testing Routine

- Weekly Testing: Perform weekly tests for pH, ammonia, nitrites, and dissolved oxygen to monitor ongoing water quality.
- Daily Checks: Monitor temperature and check for any sudden changes in water conditions daily.

ii. Record Keeping

- Documentation: Keep detailed records of all water quality tests, including dates, results, and any corrective actions taken. This helps track trends over time and provides valuable information for managing your catfish farm.

iii. Adjusting Based on Results

- Responsive Management: Use water testing data to make informed decisions about water management practices. Adjust feeding, filtration, and aeration based on the results to maintain optimal conditions.

Regular water testing is a critical component of effective catfish farming. By monitoring key water quality parameters and taking timely actions based on test results, you can ensure a healthy environment for your catfish, optimize their growth and feed efficiency, and prevent potential issues before they become serious problems. Implementing a consistent water testing routine and maintaining accurate records will help you manage water quality effectively and contribute to the overall success and sustainability of your catfish farming operation.

6.3 Common Water Quality Problems and Solutions

Maintaining high water quality is crucial for the success of your catfish farming operation. However, various water quality issues can arise, affecting the health and productivity of your fish. Understanding these common problems and knowing how to address them can help you maintain a thriving aquaculture environment. Here's an extensive guide to common water quality problems and practical solutions.

1. High Ammonia Levels

i. Problem

- Cause: Ammonia is a by product of fish waste, uneaten food, and decaying organic matter. Inadequate filtration, overfeeding, or insufficient water exchange can lead to elevated ammonia levels.

- Effects: High ammonia levels can be toxic to catfish, causing symptoms such as lethargy, gill damage, and even death. It also contributes to poor water quality and increased disease risk.

ii. Solutions

- Improve Filtration: Use high-quality biological filters to help convert ammonia into less harmful substances through nitrification. Ensure your filtration system is properly sized for your pond or tank.

- Regular Water Changes: Perform regular partial water changes to dilute ammonia concentrations and maintain water quality.

- Adjust Feeding Practices: Feed catfish smaller amounts more frequently to reduce uneaten food and subsequent ammonia production. Avoid overfeeding.

2. High Nitrite Levels

i. Problem

- Cause: Nitrites are intermediate products in the nitrogen cycle, formed from ammonia. High nitrite levels can result from inadequate biological filtration or overstocking.

- Effects: Nitrite poisoning can interfere with oxygen transport in fish blood, leading to symptoms like brown gills, gasping at the surface, and reduced growth rates.

ii. Solutions

- Enhance Filtration: Ensure that your filtration system includes both mechanical and biological components to support the conversion of nitrites into nitrates.

- Increase Aeration: Boost oxygen levels in the water to help mitigate the effects of nitrite poisoning. Use aerators or water movement devices to improve oxygenation.

- Reduce Stocking Density: Avoid overstocking to prevent excessive waste production and nitrite build up.

3. High Nitrate Levels

i. Problem

- Cause: Nitrates are the end product of the nitrification process and accumulate in the water over time. High nitrate levels can result from poor water exchange and excessive fish waste.

- Effects: While nitrates are less toxic than ammonia and nitrites, very high levels can contribute to algae blooms and poor water quality.

ii. Solutions

- Regular Water Changes: Conduct regular water exchanges to dilute nitrate concentrations and maintain water quality.

- Use Nitrate-Removing Products: Consider using nitrate-removing filters or additives designed to reduce nitrate levels in your water.

- Control Feeding and Stocking: Manage feeding practices and stocking density to minimize waste production and nitrate accumulation.

4. Low Dissolved Oxygen (DO) Levels

i. Problem

- Cause: Low DO levels can result from poor aeration, high temperatures, excessive organic matter, or overcrowding. Insufficient oxygen can stress fish and impair their health.

- Effects: Symptoms of low DO include fish gasping at the surface, reduced activity, and increased susceptibility to disease.

ii. Solutions

- Improve Aeration: Use aeration devices such as air stones, diffusers, or surface agitators to increase oxygen levels in the water. Ensure proper oxygenation, especially during warm weather or high stocking densities.

- Monitor Temperature: Maintain water temperature within the optimal range for catfish to prevent oxygen depletion. High temperatures reduce oxygen solubility in water.

- Reduce Organic Load: Minimize the build up of organic matter by managing feed and waste. Regularly clean your system and remove excess debris.

5. pH Fluctuations

i. Problem

- Cause: pH levels can fluctuate due to changes in water chemistry, such as the addition of chemicals, poor water exchange, or the accumulation of organic acids.

- Effects: Rapid or extreme pH changes can stress catfish, affecting their growth, health, and overall survival.

ii. Solutions

- Stable Buffering: Use buffering agents or additives to stabilize pH levels and prevent rapid changes. Test pH regularly to monitor stability.

- Regular Monitoring: Check pH levels frequently to detect and address fluctuations early. Maintain pH within the ideal range for catfish (usually 6.5-8.0).

- Avoid Chemical Additions: Be cautious when adding chemicals or treatments to the water. Ensure they are compatible with your system and do not cause pH imbalances.

6. Algae Blooms

i. Problem

- Cause: Algae blooms are often caused by excess nutrients (nitrogen and phosphorus) in the water, inadequate water circulation, and high light levels.

- Effects: Algae blooms can deplete oxygen levels, block light, and create a hostile environment for fish.

ii. Solutions

- Manage Nutrients: Reduce nutrient inputs by controlling feed and waste. Regularly clean your system to prevent nutrient build up.

- Improve Water Circulation: Enhance water movement to prevent stagnant areas where algae can thrive.

Understanding and addressing common water quality problems is vital for the success of your catfish farming operation. By monitoring key parameters, implementing effective solutions, and maintaining a proactive approach to water management, you can create a healthy and productive environment for your catfish. Regular testing, proper maintenance, and timely interventions will help ensure the long-term health and profitability of your farm.

CHAPTER 7

HEALTH MANAGEMENT AND DISEASE PREVENTION

7.1 Identifying Common Catfish Diseases

Identifying and managing diseases is a crucial aspect of successful catfish farming. Diseases can significantly impact the health and growth of your fish, leading to losses and reduced productivity. Understanding the signs and symptoms of common catfish diseases allows you to take timely action to prevent and control outbreaks. Here's an extensive guide to recognizing and addressing common catfish diseases.

1. Understanding Catfish Diseases

i. Importance of Early Detection

- Prevention: Early detection helps prevent the spread of diseases to other fish and minimizes potential losses.
- Treatment: Timely intervention can often cure or manage diseases more effectively, improving fish health and farm productivity.
- Management: Recognizing diseases early allows for better management of your farm, reducing the impact on your fish and operation.

ii. Common Indicators of Disease

- Behavioural Changes: Look for changes in feeding habits, swimming patterns, or lethargy.
- Physical Symptoms: Observe for visible signs such as lesions, discoloration, or unusual growths on the fish.
- Environmental Factors: Monitor water quality and changes in the environment that may contribute to disease outbreaks.

2. Common Catfish Diseases

i. Columnaris (Flexibacter Columnaris)

- Symptoms: Columnaris causes lesions or ulcers on the skin, gills, and fins. Fish may exhibit frayed fins, greyish or white patches, and difficulty breathing.
- Causes: This bacterial infection is often associated with poor water quality, high temperatures, and stress.
- Treatment: Improve water quality, use antibacterial treatments, and maintain good hygiene in the farm system. Ensure proper aeration and reduce stress factors.

ii. Ich (Ichthyophthirius multifiliis)

- Symptoms: Ich, also known as "white spot disease," presents as small white cysts or spots on the skin, gills, and fins. Fish may also show signs of scratching against surfaces and increased respiratory distress.
- Causes: This protozoan parasite thrives in stressed or overcrowded conditions and poor water quality.
- Treatment: Increase water temperature gradually to speed up the parasite's life cycle and use anti-parasitic treatments. Improve water quality and reduce stocking density.

iii. Catfish Fungus (Saprolegnia spp.)

- Symptoms: Fungus infections appear as cotton-like growths on the skin, gills, or fins. Infected fish may also exhibit lethargy and reduced feeding.
- Causes: Fungal infections often occur in fish with damaged skin or gills, typically in environments with poor water quality.
- Treatment: Improve water quality, treat with antifungal medications, and remove any sources of stress or physical damage.

iv. Dropsy

- Symptoms: Dropsy is characterized by swelling of the abdomen, scales that appear raised, and fluid accumulation. Fish may also show signs of difficulty swimming and lethargy.
- Causes: This condition is often a result of internal bacterial infections or poor water quality.
- Treatment: Address underlying causes by improving water quality and using appropriate antibiotics or anti-bacterial treatments. Ensure a balanced diet and reduce stress.

v. Swim Bladder Disease

- Symptoms: Affected fish may have difficulty maintaining buoyancy, causing them to float sideways or sink. They may also exhibit signs of abnormal swimming behaviour.
- Causes: Swim bladder disease can be caused by overfeeding, poor water quality, or physical injuries.
- Treatment: Adjust feeding practices to avoid overfeeding, improve water quality, and provide supportive care to affected fish.

3. Seeking Professional Help

i. Consulting Experts

- Veterinarians: Consult with aquaculture veterinarians or fish health experts for accurate diagnosis and treatment recommendations.
- Extension Services: Seek assistance from agricultural extension services or aquaculture associations for additional support and resources.

ii. Diagnostic Services

- Laboratory Testing: Use diagnostic services for detailed analysis and confirmation of disease pathogens. This can help in selecting the most effective treatment.

Identifying and managing catfish diseases is essential for maintaining the health and productivity of your farm. By recognizing common diseases, understanding their causes, and implementing preventive measures, you can minimize the risk of outbreaks and ensure a successful aquaculture operation. Regular monitoring, proper care, and seeking expert advice when needed will help you effectively manage fish health and improve the overall performance of your catfish farm.

7.2 Disease Prevention Strategies

Preventing diseases in catfish farming is crucial to ensure the health, growth, and productivity of your fish. Effective disease prevention strategies can help you avoid costly outbreaks and maintain a thriving aquaculture system. Here's a reader-friendly guide to implementing disease prevention strategies in your catfish farm.

1. Maintaining Optimal Water Quality

i. Importance of Water Quality

- Health Impact: Good water quality is essential for the overall health and well-being of catfish. Poor water quality can stress fish, weaken their immune system, and make them more susceptible to diseases.

- Preventive Measures: Regular monitoring and maintenance of water quality can prevent many disease issues.

ii. Parameters to Monitor

- Temperature: Maintain water temperature within the optimal range for your catfish species. Extreme temperatures can stress fish and lead to disease.

- pH Level: Ensure pH levels remain stable and within the recommended range for catfish (usually between 6.5 and 8.0).

- Ammonia, Nitrite, and Nitrate Levels: Regularly test and manage these parameters to prevent toxicity and maintain a healthy environment.

iii. Practical Tips

- Filtration: Use high-quality filtration systems to remove waste and maintain water clarity.

- Aeration: Provide adequate aeration to ensure sufficient oxygen levels and prevent stagnation.

- Regular Cleaning: Clean tanks, ponds, and equipment regularly to reduce the build up of organic matter and pathogens.

2. Implementing Proper Feeding Practices

i. Balanced Diet

- Nutritional Needs: Provide a well-balanced diet that meets the nutritional requirements of your catfish. Proper nutrition supports immune function and overall health.
- Quality Feed: Use high-quality feed that is specifically formulated for catfish to ensure optimal growth and health.

ii. Feeding Techniques

- Avoid Overfeeding: Feed catfish appropriate amounts to prevent uneaten food from decomposing and affecting water quality.
- Feeding Schedule: Establish a consistent feeding schedule to promote healthy growth and minimize waste.

iii. Practical Tips

- Monitor Intake: Observe feeding behaviour and adjust the amount of feed as needed to match the fish's appetite and growth stage.
- Store Feed Properly: Store feed in a dry, cool place to prevent spoilage and contamination.

3. Managing Stocking Density

i. Importance of Stocking Density

- Disease Spread: Overcrowding can lead to stress, competition for resources, and increased disease risk. Maintaining appropriate stocking density helps reduce the likelihood of disease outbreaks.
- Growth and Health: Proper stocking density ensures that fish have adequate space to grow and thrive.

ii. Guidelines for Stocking Density

- Space Requirements: Follow recommended space requirements for the specific catfish species you are farming. This can vary based on the size of the fish and the type of farming system.

- Monitoring: Regularly assess fish growth and adjust stocking density as needed to maintain a healthy environment.

iii. Practical Tips

- Gradual Increase: Introduce fish gradually to avoid sudden increases in stocking density that could overwhelm the system.
- Assess Health: Regularly check fish for signs of stress or disease and adjust stocking density if needed.

4. Implementing Biosecurity Measures

i. Importance of Biosecurity

- Disease Prevention: Biosecurity measures help prevent the introduction and spread of diseases within your farm. They are essential for maintaining a healthy and productive fish population.
- Protecting Assets: Proper biosecurity can protect your investment and ensure long-term success.

ii. Biosecurity Practices

- Quarantine: Isolate new fish before introducing them to your main population. This helps prevent the spread of potential diseases.
- Disinfection: Regularly disinfect equipment, tanks, and ponds to kill pathogens and reduce the risk of contamination.
- Controlled Access: Restrict access to your farm to prevent the introduction of diseases from external sources.

iii. Practical Tips

- Footbaths and Hand Sanitizers: Use footbaths and hand sanitizers for anyone entering the farm to minimize the risk of introducing pathogens.

- Regular Inspections: Conduct regular inspections of your farm and equipment to identify and address potential biosecurity issues.

5. Stress Reduction Strategies

i. Importance of Reducing Stress

- Immune Function: Stress can weaken the immune system of catfish, making them more susceptible to diseases. Reducing stress helps maintain their health and resilience.

- Disease Prevention: Minimizing stress factors can prevent the onset of stress-related diseases and improve overall farm performance.

ii. Stress Management Techniques

- Consistent Environment: Maintain stable water conditions and avoid sudden changes in temperature, pH, or other parameters.

- Gentle Handling: Handle fish carefully and minimize disturbances to reduce stress.

- Adequate Space: Provide sufficient space and resources to prevent overcrowding and competition.

iii. Practical Tips

- Monitor Behaviour: Observe fish behaviour for signs of stress, such as unusual swimming patterns or reduced feeding, and take corrective actions as needed.

- Provide Hiding Places: Incorporate structures or hiding places in tanks or ponds to allow fish to escape from perceived threats and reduce stress.

6. Regular Health Monitoring

i. Importance of Monitoring

- Early Detection: Regular health monitoring helps detect diseases early, allowing for prompt intervention and treatment.

- Overall Health: Routine checks ensure that fish remain healthy and that any potential issues are addressed before they become serious problems.

ii. Monitoring Techniques

- Visual Inspections: Conduct daily visual inspections of your fish for signs of illness, injury, or abnormal behaviour.
- Health Records: Maintain detailed health records to track fish health, treatments, and any incidents of disease.

iii. Practical Tips

- Observation: Pay close attention to changes in fish behaviours, appearance, and feeding habits.
- Expert Consultation: Consult with aquaculture professionals or veterinarians if you notice any signs of illness or if you need guidance on health management.

Effective disease prevention is essential for the success of your catfish farming operation. By maintaining optimal water quality, implementing proper feeding practices, managing stocking density, and following biosecurity measures, you can significantly reduce the risk of disease outbreaks. Regular health monitoring, stress reduction, and ongoing education further contribute to a healthy and productive aquaculture environment. With these strategies in place, you can achieve a thriving catfish farm and ensure the long-term success of your aquaculture endeavours.

7.3 Treatment and Recovery

When diseases strike your catfish farm, prompt and effective treatment is crucial to restoring health and ensuring the continued success of your operation. Understanding the treatment options and recovery strategies will help you manage outbreaks efficiently and support your fish through their healing process. Here's a guide on how to approach treatment and recovery for your catfish.

1. Identifying the Disease

i. Importance of Accurate Diagnosis

- Effective Treatment: Accurate diagnosis ensures that you select the most appropriate treatment for the specific disease affecting your fish.
- Prevention of Resistance: Correct diagnosis helps prevent the misuse of medications, which can lead to drug resistance and treatment failures.

ii. Methods for Diagnosis

- Visual Inspection: Examine fish for visible symptoms such as lesions, spots, or abnormal behaviours.
- Diagnostic Testing: Use laboratory tests to identify pathogens and confirm the disease. This may include water quality tests, bacterial cultures, or microscopic examination.

iii. Consulting Experts

- Veterinarians: Consult with aquaculture veterinarians for professional diagnosis and treatment recommendations.
- Extension Services: Reach out to local agricultural or aquaculture extension services for advice and support.

2. Treatment Options

i. Medication

- Antibiotics: Used to treat bacterial infections. Follow dosage instructions carefully and complete the full course of treatment to prevent relapse.
- Antifungals: Applied to address fungal infections. Ensure proper application and dosage to effectively eliminate the fungus.
- Antiparasitics: Target external or internal parasites. Choose treatments based on the specific type of parasite and follow application guidelines.

ii. Non-Medication Treatments

- Salt Baths: Salt can be used to treat external parasites and fungal infections. Dissolve non-iodized salt in water and immerse affected fish for a specified duration.
- Temperature Adjustment: Gradually adjusting water temperature can help manage certain diseases, such as ich. Ensure changes are gradual to avoid stressing the fish.

iii. Water Quality Management

- Improving Water Conditions: Addressing issues like poor water quality can help alleviate stress and promote healing. Regularly monitor and adjust parameters such as pH, temperature, and ammonia levels.
- Filtration and Aeration: Enhance filtration and aeration to maintain a clean and well-oxygenated environment, supporting fish recovery.

3. Implementing Treatment

i. Preparing for Treatment

- Isolate Affected Fish: Quarantine infected fish to prevent the spread of disease to healthy fish. Use separate tanks or sections for treatment.
- Follow Instructions: Adhere to the manufacturer's instructions for medications and treatments. Overdosing or improper use can be harmful to fish and ineffective against the disease.

ii. Administering Treatment

- Dosage: Accurately measure and administer the correct dosage of medication. Avoid overuse or underuse, as it can impact treatment effectiveness and fish health.
- Monitoring: Observe fish closely during treatment for any adverse reactions or changes in symptoms. Adjust treatment if necessary based on their response.

iii. Post-Treatment Care

- Continue Monitoring: Continue to monitor the health of your fish after treatment to ensure recovery and detect any signs of relapse or secondary infections.
- Gradual Reintroduction: Gradually reintroduce treated fish into the main population once they have fully recovered and the water quality is stable.

4. Recovery and Support

i. Supporting Recovery

- Nutrition: Provide a balanced and nutritious diet to support the recovery of your fish. Proper nutrition helps strengthen the immune system and promote healing.
- Stress Reduction: Minimize stress by maintaining stable water conditions and avoiding sudden changes in the environment.

ii. Monitoring and Follow-Up

- Regular Health Checks: Conduct regular health checks to ensure that fish remain healthy and are free from disease.
- Record Keeping: Maintain detailed records of treatments, observations, and recovery progress for future reference and management.

iii. Prevention of Recurrence

- Educational Resources: Stay informed about best practices for disease prevention and management to continuously improve your farming techniques.

Effective treatment and recovery strategies are essential for managing diseases in your catfish farm. By accurately diagnosing diseases, selecting appropriate treatments, and providing ongoing support for recovery, you can improve the health and productivity of your fish. Implementing preventive measures and maintaining good farm management practices will further safeguard against future outbreaks, ensuring a successful and thriving catfish farming operation.

CHAPTER 8
HARVESTING AND MARKETING

8.1 When to Harvest Your Catfish

Harvesting your catfish at the right time is crucial to ensure that you get the best quality fish and maximize the profitability of your farm. Knowing when to harvest involves understanding the growth stages of your catfish, monitoring their health and size, and making informed decisions based on market demand and farming conditions. Here's an extensive, reader-friendly guide on when to harvest your catfish.

1. Understanding Catfish Growth Stages

i. Early Growth Stages

- Fingerlings: These are young catfish that are not yet ready for market. They are usually between 2 to 6 inches long and need further growth before harvesting.

- Juveniles: Catfish in this stage are growing rapidly but are still not fully mature. They are generally between 6 to 12 inches long.

ii. Market Size

- Grow-Out Stage: This is the stage where catfish are growing towards their market size. The length and weight of the fish will depend on the specific breed and farming practices. Typically, market-size catfish are between 12 to 24 inches long and weigh 1 to 3 pounds.

iii. Maturity

- Fully Mature: Mature catfish are fully grown and have reached their maximum weight and size. They are often larger than market size but may not always be the ideal size for sale.

2. Factors Influencing Harvest Timing

i. Desired Market Size

- Size Requirements: Different markets and buyers may have specific size requirements for catfish. Research local market preferences and target the size that aligns with demand.

- Growth Rate: Consider the growth rate of your catfish breed and how long it takes to reach the desired size. Adjust your feeding and management practices to optimize growth.

ii. Health and Condition of Fish

- Health Monitoring: Regularly check the health of your catfish. Fish that are healthy and in good condition are more likely to meet market standards.
- Disease or Stress: If fish are showing signs of disease or stress, it may be better to harvest them early to prevent further issues and avoid compromising the quality of your harvest.

iii. Environmental Conditions

- Water Quality: Ensure that water quality is optimal for growth. Poor water conditions can affect the health and growth of your fish, influencing the timing of harvest.
- Seasonal Factors: Weather and seasonal changes can impact fish growth rates. Plan your harvest timing to align with stable conditions for optimal results.

3. Harvesting Techniques

i. Methods of Harvesting

- Netting: Use large nets to capture catfish from ponds or tanks. This method is common for small to medium-sized operations.
- Trapping: Employ traps to catch fish, especially in larger bodies of water. This can be effective for managing larger populations.
- Pumping: For commercial-scale operations, consider using pumps to remove water and fish from tanks or ponds efficiently.

ii. Handling and Processing

- Gentle Handling: Handle fish gently to avoid stress and injury. Use soft nets or slings to move them without causing damage.
- Processing: Once harvested, fish should be processed promptly to maintain freshness. This includes cleaning, filleting, and packaging according to market standards.

4. Timing for Optimal Harvest

i. Economic Considerations

- Profitability: Calculate the cost of feeding and maintaining fish versus the potential selling price. Harvesting at the optimal size and weight can maximize profitability.
- Market Demand: Align your harvest timing with market demand to ensure that you can sell your fish at a good price. Monitor local markets and adjust your harvest schedule accordingly.

ii. Growth Potential

- Growth Trends: Monitor growth trends and adjust your feeding and management practices to achieve the desired size at the right time.
- Harvest Schedule: Develop a harvest schedule that allows for consistent production and sales. This may involve staggering harvests to ensure a steady supply of fish.

5. Preparing for Harvest

i. Planning and Coordination

- Harvest Plan: Develop a detailed harvest plan, including the timing, methods, and equipment needed. Ensure that all necessary resources are in place before starting the process.
- Staff Training: Train staff on harvesting techniques, handling procedures, and safety protocols to ensure an efficient and effective harvest.

ii. Post-Harvest Management

- Cleaning and Maintenance: After harvesting, clean and maintain tanks, ponds, and equipment to prepare for the next cycle of fish.

- Record Keeping: Keep detailed records of harvest dates, fish sizes, and other relevant data to track performance and inform future planning.

Knowing when to harvest your catfish involves a combination of understanding growth stages, monitoring health and environmental conditions, and aligning with market demands. By carefully considering these factors, you can determine the optimal timing for harvesting your fish, ensuring high quality and maximizing profitability. Implement effective harvesting techniques and planning to achieve the best results for your catfish farm and ensure a successful operation.

8.2 Harvesting Techniques

Harvesting catfish efficiently and humanely is key to maintaining the quality of your product and the overall health of your farm. Different harvesting techniques can be used depending on the size of your operation, the type of farming system, and the scale of your harvest. Here's an extensive, reader-friendly guide on various catfish harvesting techniques.

1. Netting

i. Hand Nets

- Description: Hand nets are small, handheld nets used for catching individual fish or small quantities.

- Best For: This method is ideal for smaller ponds or tanks and for harvesting fish in manageable quantities.

- Procedure: Gently scoop the net through the water, ensuring minimal stress to the fish. Transfer the fish to a container or holding area.

ii. Seine Nets

- Description: Seine nets are large nets with floats on the top and weights on the bottom. They are used to enclose and capture groups of fish.

- Best For: Suitable for larger ponds or tanks where fish are more numerous.
- Procedure: Deploy the net from a boat or the shore, dragging it through the water to encircle the fish. Once enclosed, pull the net to the shore or boat and gather the fish.

iii. Gill Nets

- Description: Gill nets are vertical nets with mesh sizes that catch fish by their gills.
- Best For: Effective in larger bodies of water or for harvesting specific sizes of fish.
- Procedure: Set up the net vertically in the water where fish are known to swim. Check the net regularly and remove fish that are caught.

2. Trapping

i. Fish Traps

- Description: Fish traps are containers designed to attract and capture fish. They come in various designs, such as funnel-shaped traps.
- Best For: Ideal for capturing fish in ponds or tanks with minimal disturbance.
- Procedure: Place traps in areas where fish are most active. Check the traps regularly and remove captured fish.

ii. Bait Traps

- Description: Bait traps use bait to attract fish into a confined space where they are then trapped.
- Best For: Useful for catching fish when their movement is restricted or when targeting specific fish sizes.
- Procedure: Bait the trap with attractants and place it in the water. Monitor the trap and collect fish as needed.

3. Pumping

i. Mechanical Pumps

- Description: Mechanical pumps are used to remove water and fish from tanks or ponds in bulk.
- Best For: Large-scale operations where efficiency and speed are crucial.
- Procedure: Set up the pump to drain water from the farming system. The fish will be collected in a net or container as the water is removed.

ii. Vacuum Pumps

- Description: Vacuum pumps use suction to remove fish and water from tanks or ponds.
- Best For: Useful in aquaculture systems with controlled environments.
- Procedure: Use the vacuum pump to suction water and fish from the tank. Transfer the fish to a holding area for processing.

4. Harvesting from Tanks

i. Draining and Sorting

- Description: This method involves draining the tank and then sorting the fish.
- Best For: Efficient for tanks with a manageable volume of water and fish.
- Procedure: Gradually drain the water while ensuring that fish are captured using nets. Sort the fish by size and condition as they are removed.

ii. Use of Fish Graders

- Description: Fish graders are mechanical devices that sort fish by size as they are harvested.

- Best For: Automated systems that require sorting by size for market or processing needs.
- Procedure: Pass the fish through the grader, which separates them into different size categories. Collect fish in separate bins or containers based on size.

5. Harvesting from Ponds

i. Drainage

- Description: Drainage involves lowering the water level in the pond to facilitate easier capture of fish.
- Best For: Useful for large ponds where other methods may be less effective.
- Procedure: Slowly drain the pond to concentrate the fish. Use nets or traps to collect the fish as the water level decreases.

ii. Harvesting Cages

- Description: Cages are used to confine fish within a specific area of the pond for easier collection.
- Best For: Effective for managing fish populations and simplifying harvest.
- Procedure: Place the cages in the pond and allow fish to accumulate inside. Remove the cage and collect the fish for processing.

6. Handling and Processing

i. Gentle Handling

- Description: Proper handling techniques reduce stress and injury to fish.
- Best For: Ensuring high-quality fish and maintaining their health.
- Procedure: Use soft nets or slings to handle fish. Avoid rough handling and minimize the time fish spend out of the water.

ii. Immediate Processing

- Description: Processing fish promptly after harvest preserves freshness and quality.

- Best For: Ensuring that fish meet market standards and are ready for sale.

- Procedure: Clean, fillet, and package fish according to market requirements. Store processed fish under appropriate conditions to maintain quality.

Choosing the right harvesting technique depends on the size of your operation, the type of farming system, and the scale of your harvest. Whether using nets, traps, pumps, or cages, the goal is to efficiently and humanely capture your catfish while minimizing stress and ensuring high quality. By implementing effective harvesting methods and handling practices, you can achieve a successful harvest and maintain the overall health and profitability of your catfish farm.

8.3 Selling Your Catfish

Selling your catfish is the final step in the farming process, and it's crucial for ensuring that you achieve profitability and meet market demands. The process involves several key steps, from preparing your fish for sale to finding the right buyers and managing sales effectively. Here's an in-depth guide on how to sell your catfish successfully.

1. Preparing Your Catfish for Sale

i. Cleaning and Processing

- Cleaning: Ensure that all catfish are cleaned thoroughly to remove any residual dirt or contaminants. This includes washing and rinsing the fish under clean, running water.

- Processing: Depending on market requirements, you may need to fillet or gut the fish. Processing should be done quickly and efficiently to maintain freshness and quality.

- Packaging: Package fish in appropriate materials to keep them fresh. Common packaging includes vacuum-sealed bags or ice-filled containers. Label packages with important information such as weight, date, and handling instructions.

ii. Quality Control

- Inspection: Inspect the fish for any signs of disease or damage. Only sell fish that meet high-quality standards.
- Temperature Management: Keep the fish at the proper temperature to ensure they remain fresh until they reach the market. Use ice or refrigeration as needed.

2. Finding Buyers

i. Local Markets

- Fish Markets: Local fish markets are a common place to sell catfish. Establish relationships with market vendors and understand their requirements for fish quality and size.
- Restaurants: Many restaurants seek fresh, locally-sourced catfish. Reach out to local dining establishments and offer samples or product information.
- Grocery Stores: Grocery stores may be interested in stocking fresh catfish. Contact store managers and provide details about your farm and product.

ii. Wholesale and Distribution

- Distributors: Wholesale distributors can help you reach a broader market. They often have established networks and can manage large orders.
- Online Platforms: Consider using online platforms to sell your catfish. Websites and apps dedicated to local food sales can connect you with buyers directly.
- Cooperatives: Join a cooperative or farmer's association that specializes in fish farming. These groups often provide marketing support and collective selling opportunities.

iii. Export Opportunities

- International Markets: If you're interested in exporting, research international markets where there is demand for catfish. Comply with export regulations and quality standards for different countries.

3. Marketing Your Catfish

i. Branding

- Farm Branding: Develop a brand for your catfish farm. This includes creating a logo, farm name, and promotional materials that highlight the quality and unique aspects of your product.
- Quality Assurance: Promote the freshness and quality of your catfish through your branding. Emphasize any certifications or sustainable practices that differentiate your farm.

ii. Promotional Strategies

- Sampling: Offer samples of your catfish to potential buyers to showcase the quality and taste.
- Advertising: Use local newspapers, online ads, and social media to advertise your catfish. Include attractive images and details about your farm and products.
- Events: Participate in local food festivals, farmers' markets, or trade shows to promote your catfish and connect with potential customers.

4. Managing Sales

i. Pricing

- Market Research: Research local and regional prices for catfish to set competitive prices. Consider factors such as size, quality, and packaging.
- Cost Analysis: Calculate your costs for farming, processing, and distribution to ensure your pricing covers expenses and provides a profit margin.

ii. Sales Channels

- Direct Sales: Selling directly to consumers or local businesses allows you to retain a higher profit margin. Build relationships with buyers and offer personalized service.

- Bulk Sales: For larger quantities, consider selling in bulk to wholesalers or distributors. This may require less frequent sales but can provide steady revenue.

iii. Customer Service

- Communication: Maintain clear communication with buyers regarding orders, delivery times, and any issues that arise.

- Feedback: Collect feedback from customers to improve your product and service. Address any concerns promptly to build a positive reputation.

Successfully selling your catfish involves thorough preparation, effective marketing, and efficient sales management. By focusing on quality, building strong relationships with buyers, and understanding market demands, you can achieve a profitable and sustainable business. Continuously assess your selling strategies, stay informed about market trends, and adapt to changes to ensure ongoing success in your catfish farming venture.

CHAPTER 9

FINANCIAL MANAGEMENT AND SUSTAINABILITY

9.1 Budgeting for Your Catfish Farm

Budgeting is a critical step in establishing and running a successful catfish farm. It involves planning and managing financial resources to ensure that your farm operates efficiently and remains profitable. A well-thought-out budget helps you understand your start up costs, manage ongoing expenses, and plan for future investments. Here's a straightforward guide on budgeting for your catfish farm.

1. Understanding Start up Costs

i. Initial Investment

- Land and Infrastructure: Costs associated with purchasing or leasing land and developing the necessary infrastructure, such as ponds, tanks, and buildings. This may include grading, construction, and installation of water systems.

- Equipment: Investment in essential equipment such as pumps, aerators, nets, feeders, and water quality testing tools.

- Fingerlings: Initial purchase of catfish fingerlings to start your farming operation. Prices can vary based on the breed and quantity.

- Permits and Licenses: Fees for obtaining necessary permits and licenses for operating a catfish farm, which can include environmental permits, health and safety certifications, and business licenses.

- Operating Capital: Funds needed to cover initial operational expenses until your farm becomes self-sustaining. This includes costs for feed, utilities, and labour.

ii. Contingency Fund

- Unexpected Expenses: Set aside a portion of your budget for unexpected costs, such as repairs, equipment failures, or emergencies. A common practice is to allocate 10-20% of your total budget for contingencies.

2. Calculating Operating Costs

i. Feed and Nutrition

- Feed Costs: Regular expenses for purchasing catfish feed, which is a significant part of your ongoing costs. Costs depend on the type of feed, quality, and the number of fish being fed.
- Supplementary Nutrition: Additional costs for vitamins, minerals, and other supplements that may be needed to ensure optimal growth and health of your fish.

ii. Labour Costs

- Wages: Costs associated with hiring and compensating farm workers. This includes salaries, benefits, and any additional expenses related to labour.
- Training: Investment in training programs or resources for employees to ensure they are knowledgeable about catfish farming practices.

iii. Utilities and Maintenance

- Water Supply: Expenses for water usage, including treatment and filtration systems.
- Electricity: Costs for operating equipment such as aerators, pumps, and lighting.
- Maintenance: Regular maintenance costs for equipment and infrastructure to ensure everything operates efficiently and to prevent breakdowns.

3. Financial Projections and Planning

i. Revenue Forecasting

- Sales Estimates: Project potential income based on expected harvests, market prices, and sales volumes.

Consider different scenarios (e.g., best case, worst case) to account for market fluctuations.

- Pricing Strategy: Develop a pricing strategy that covers your costs and ensures a profit margin. Consider factors such as market demand, competition, and quality of your catfish.

ii. Cash Flow Management

- Cash Flow Statements: Regularly prepare cash flow statements to track incoming and outgoing cash. This helps ensure that you have enough liquidity to cover operating expenses and unexpected costs.

- Break-Even Analysis: Calculate your break-even point, which is the level of sales needed to cover your costs. This helps you understand how much you need to earn to start making a profit.

4. Funding and Financing Options

i. Personal Savings

- Using Savings: Assess your personal savings as a potential source of initial funding. This can be a cost-effective way to finance your start up without incurring debt.

ii. Loans and Grants

- Bank Loans: Explore options for agricultural loans from banks or financial institutions. These loans often have specific terms for farm businesses.

- Grants: Look for government or private grants that support agricultural development and fish farming. These can provide non-repayable funds to help with start up costs.

iii. Investor Funding

- Seeking Investors: Consider attracting investors who are interested in funding your catfish farm. Be prepared to present a detailed business plan and financial projections to potential investors.

iv. Partnerships

- Joint Ventures: Form partnerships with other businesses or individuals who have an interest in fish farming. This can help share costs and risks associated with starting and running a farm.

5. Monitoring and Adjusting Your Budget

i. Regular Review

- Budget Monitoring: Regularly review your budget and compare actual expenses to your projections. This helps you stay on track and identify any discrepancies.
- Adjustments: Make adjustments to your budget as needed based on actual performance and changes in costs or revenues. Flexibility is key to managing financial challenges and opportunities.

ii. Financial Reports

- Tracking Performance: Generate financial reports to track the performance of your farm. This includes income statements, balance sheets, and cash flow statements.
- Analysis: Analyse financial data to assess profitability, identify areas for improvement, and make informed decisions about future investments or adjustments.

Budgeting for your catfish farm is essential for managing costs, ensuring profitability, and planning for future growth. By carefully estimating start up costs, calculating operating expenses, and developing a solid financial plan, you can set your farm up for success. Regular monitoring and adjustment of your budget will help you navigate financial challenges and make informed decisions, ultimately contributing to the long-term sustainability and profitability of your catfish farming venture.

9.2 Managing Your Profit Margins

Managing profit margins is crucial for the success and sustainability of your catfish farm. It involves controlling costs, maximizing revenues, and ensuring that your operations are financially efficient. By effectively managing profit margins, you can enhance your farm's profitability and ensure long-term success. Here's guide to managing your profit margins in catfish farming.

1. Understanding Profit Margins

What Are Profit Margins?

Profit margins represent the percentage of revenue that remains after all costs and expenses have been deducted. It's a key indicator of financial health and efficiency.

Types of Margins:

- Gross Profit Margin: Calculated as (Revenue - Cost of Goods Sold) / Revenue. It reflects the profitability of your core operations before accounting for overheads.
- Net Profit Margin: Calculated as (Net Income / Revenue). It shows the overall profitability of your business after all expenses, including taxes and interest, have been deducted.

Importance of Profit Margins

- Financial Health: Higher profit margins indicate better financial health and efficiency in managing costs relative to revenue.
- Sustainability: Managing profit margins effectively ensures that your farm can withstand market fluctuations and economic challenges.

2. Controlling Costs

i. Identifying Major Costs

- Feed Costs: The largest ongoing expense in catfish farming. Optimize feed usage by selecting high-quality, cost-effective feed and adjusting feeding rates based on fish growth stages.
- Labour Costs: Expenses related to hiring and compensating workers. Improve efficiency by training staff, streamlining tasks, and employing labour-saving technologies.

- Utilities: Costs for water, electricity, and other utilities. Reduce these costs through energy-efficient equipment and water conservation practices.

ii. Reducing Expenses

- Bulk Purchasing: Buy feed, equipment, and supplies in bulk to benefit from discounts and reduce per-unit costs.
- Maintenance: Implement regular maintenance schedules to prevent costly breakdowns and extend the lifespan of your equipment.
- Waste Management: Minimize waste by optimizing feed use and recycling or repurposing by-products.

iii. Budget Monitoring

- Tracking Expenses: Keep detailed records of all expenses to identify areas where costs can be reduced.
- Regular Reviews: Regularly review your budget and compare it to actual expenses. Adjust your budget and cost-control measures as needed.

3. Maximizing Revenue

i. Enhancing Sales

- Quality: Ensure high quality of your catfish to attract premium prices. Focus on fresh, well-sized, and healthy fish.
- Marketing: Develop effective marketing strategies to reach a broader audience. Utilize social media, local advertising, and direct sales approaches.
- Diversification: Explore different sales channels, such as local markets, restaurants, and online platforms, to increase your revenue streams.

ii. Pricing Strategy

- Competitive Pricing: Set competitive prices based on market research and your cost structure. Consider factors such as market demand, competition, and fish quality.

- Value Addition: Offer value-added products, such as fillets or processed catfish, to increase your revenue per unit.

iii. Customer Relationships

- Loyalty Programs: Implement loyalty programs or incentives for repeat customers to encourage continued business.
- Feedback: Gather and act on customer feedback to improve your products and services, which can lead to increased sales and customer satisfaction.

4. Improving Operational Efficiency

i. Streamlining Operations

- Process Optimization: Identify and eliminate inefficiencies in your farming processes. Streamline tasks such as feeding, harvesting, and water management.
- Technology Integration: Invest in technology that improves efficiency, such as automated feeders, water quality monitoring systems, and data management tools.

ii. Staff Training

- Training Programs: Provide training to your staff to ensure they are skilled and efficient in their roles. This can lead to improved productivity and reduced errors.
- Motivation: Implement incentive programs to motivate employees and enhance their performance.

5. Financial Management

i. Financial Planning

- Projections: Develop financial projections to forecast revenues, expenses, and profit margins. Use these projections to guide decision-making and adjust strategies.

- Cash Flow Management: Monitor cash flow to ensure you have sufficient liquidity to cover expenses and take advantage of growth opportunities.

ii. Profit Analysis

- Margin Analysis: Regularly analyse your profit margins to assess financial performance. Identify trends and areas for improvement.
- Cost-Benefit Analysis: Evaluate the cost versus benefit of investments or changes in operations to ensure they contribute positively to your profit margins.

Effectively managing your profit margins is essential for the success of your catfish farm. By controlling costs, maximizing revenue, improving operational efficiency, and staying adaptable to market changes, you can enhance your profitability and ensure the long-term viability of your farm. Regularly review and adjust your financial strategies to maintain healthy profit margins and achieve sustainable growth.

9.3 Sustainable Catfish Farming Practices

Sustainable catfish farming practices are essential for minimizing environmental impact, conserving resources, and ensuring the long-term viability of your farming operation. These practices focus on balancing productivity with environmental stewardship, social responsibility, and economic viability. Here's a reader-friendly guide to sustainable catfish farming practices.

1. Understanding Sustainability in Catfish Farming

i. What is Sustainable Farming?

- Definition: Sustainable farming refers to practices that meet current needs without compromising the ability of future generations to meet their own needs. It involves managing resources responsibly, minimizing environmental impact, and promoting social and economic well-being.

- Importance: Adopting sustainable practices helps protect natural resources, reduces pollution, and supports the long-term health of ecosystems and communities.

ii. Benefits of Sustainable Practices

- Environmental Protection: Reduces pollution and conserves natural resources.
- Economic Efficiency: Improves resource use efficiency, potentially reducing costs and increasing profitability.
- Social Responsibility: Enhances community relations and supports fair labour practices.

2. Water Management

i. Efficient Water Use

- Water Conservation: Implement water-saving techniques such as recirculating systems and rainwater harvesting to reduce water consumption.
- Efficient Systems: Use efficient water filtration and circulation systems to minimize water wastage and maintain high water quality.

ii. Pollution Control

- Waste Management: Properly manage and treat waste to prevent pollution. Use waste products as fertilizer or in other beneficial applications.
- Avoid Overfeeding: Prevent overfeeding to reduce the accumulation of uneaten feed and minimize waste.

iii. Pond and Tank Management

- Natural Filtering: Design ponds and tanks to incorporate natural filtration systems, such as aquatic plants and biofilters, to maintain water quality and reduce the need for chemical treatments.
- Regular Monitoring: Continuously monitor water quality parameters and adjust management practices as needed

to prevent pollution and maintain a healthy environment.

3. Feed Management

i. Sustainable Feed Options

- Alternative Feeds: Explore alternative feed sources that are more sustainable, such as insect larvae, algae, or plant-based feeds, which can reduce reliance on traditional fishmeal and soybean-based feeds.
- Efficient Feed Conversion: Choose high-quality feeds that improve feed conversion rates, reducing waste and enhancing growth efficiency.

ii. Reducing Feed Waste

- Precise Feeding: Use automated feeders and monitoring systems to ensure precise feeding and minimize feed waste.
- Feeding Practices: Adjust feeding practices based on fish growth stages and environmental conditions to optimize feed use and reduce excess feed.

4. Disease Management

i. Preventive Measures

- Biosecurity: Implement strict biosecurity measures to prevent the introduction and spread of diseases. This includes controlling access to your farm, disinfecting equipment, and managing wildlife interactions.
- Health Monitoring: Regularly monitor fish health and provide timely interventions to prevent disease outbreaks and reduce the need for antibiotics.

ii. Treatment Alternatives

- Integrated Pest Management (IPM): Use IPM strategies to manage pests and diseases in an environmentally friendly way. This may include biological controls, such as introducing natural predators, and reducing chemical use.

- Natural Remedies: Explore natural or organic treatments for common fish diseases to minimize the impact on the environment and reduce chemical residues.

5. Energy Efficiency

i. Reducing Energy Consumption

- Energy-Efficient Equipment: Invest in energy-efficient equipment and technologies, such as LED lighting and energy-saving pumps, to reduce energy consumption and operational costs.
- Renewable Energy: Consider using renewable energy sources, such as solar or wind power, to reduce reliance on fossil fuels and decrease your farm's carbon footprint.

ii. Conservation Practices

- Energy Audits: Regularly conduct energy audits to identify areas for improvement and implement energy-saving measures.
- Process Optimization: Optimize operational processes to reduce energy use and enhance overall efficiency.

6. Community and Labour Practices

i. Fair Labour Practices

- Employee Welfare: Ensure fair wages, safe working conditions, and opportunities for training and advancement for your employees.
- Community Engagement: Engage with local communities and support social initiatives that contribute to the well-being of the community.

ii. Education and Training

- Sustainability Training: Provide training for your staff on sustainable farming practices and the importance of environmental stewardship.

- Knowledge Sharing: Share knowledge and best practices with other farmers and stakeholders to promote sustainable farming within the industry.

7. Certification and Standards

7.1 Sustainable Certifications

- Certifications: Obtain certifications from recognized sustainable aquaculture organizations, such as the Aquaculture Stewardship Council (ASC) or Global G.A.P. These certifications demonstrate your commitment to sustainable practices and can enhance marketability.

7.2 Adherence to Standards

- Industry Standards: Adhere to industry standards and guidelines for sustainable aquaculture to ensure compliance and maintain best practices.

Adopting sustainable catfish farming practices is essential for minimizing environmental impact, conserving resources, and ensuring the long-term success of your farm. By focusing on efficient water management, sustainable feed options, energy conservation, fair labour practices, and continuous improvement, you can achieve a balance between productivity and environmental stewardship. Embracing these practices not only enhances the sustainability of your farm but also contributes to the overall health and well-being of the ecosystem and community.

CONCLUSION

Challenges to Expect in Catfish Farming and How to Overcome Them

Starting and operating a catfish farm comes with its own set of challenges. Being prepared for these challenges and having strategies in place to address them can significantly improve the success of your farm. Here's a guide to the common challenges you might face and how to overcome them.

1. Water Quality Issues

Challenges:

- Contaminants: Presence of pollutants, such as heavy metals, pesticides, or excess nutrients, can affect fish health and growth.

- Imbalances: Fluctuations in pH, oxygen levels, and ammonia can lead to stress or disease.

Solutions:

- Regular Testing: Frequently monitor water quality parameters using reliable testing equipment.

- Filtration Systems: Invest in high-quality filtration systems to remove contaminants and maintain balanced water conditions.

- Water Treatments: Use appropriate treatments for adjusting pH and ammonia levels, and employ aeration systems to ensure adequate oxygen levels.

2. Disease Management

Challenges:

- Disease Outbreaks: Catfish are susceptible to various diseases, which can spread quickly and cause significant losses.
- Treatment Costs: Managing diseases often requires costly treatments and medications.

Solutions:

- Biosecurity Measures: Implement strict biosecurity protocols to prevent the introduction of diseases. This includes disinfecting equipment and controlling farm access.
- Health Monitoring: Regularly inspect fish for signs of illness and maintain good overall hygiene in your farming systems.

3. Feed Management

Challenges:

- High Feed Costs: Feed can be one of the largest expenses in catfish farming.
- Feed Waste: Inefficient feeding practices can lead to wasted feed and increased costs.

Solutions:

- Efficient Feed Use: Use high-quality, nutritionally balanced feeds that improve feed conversion rates and reduce waste.
- Feeding Practices: Implement precise feeding schedules and monitor fish growth to adjust feed quantities accordingly.
- Alternative Feeds: Explore alternative feed sources, such as plant-based or insect-based feeds, which may be more cost-effective and sustainable.

4. Environmental Concerns

Challenges:

- Pollution: Improper waste management can lead to pollution of surrounding environments.
- Resource Use: Overuse of water and energy can impact local resources and increase operational costs.

Solutions:

- Waste Management: Develop a waste management plan that includes recycling and treating waste products. Use organic waste as fertilizer or for other beneficial purposes.
- Resource Efficiency: Implement water-saving technologies and energy-efficient systems to minimize resource use and environmental impact.

5. Regulatory Compliance

Challenges:

- Changing Regulations: Regulations regarding aquaculture can change, requiring constant attention to ensure compliance.
- Permit Requirements: Obtaining and maintaining the necessary permits and licenses can be complex and time-consuming.

Solutions:

- Stay Informed: Regularly review local and national regulations related to catfish farming to ensure compliance.
- Consult Experts: Work with legal and regulatory experts to navigate the requirements and maintain up-to-date permits.

6. Market Fluctuations

Challenges:

- Price Volatility: Market prices for catfish can fluctuate due to supply and demand changes, impacting profitability.
- Demand Variability: Consumer preferences and market demand can shift, affecting sales and revenue.

Solutions:

- Market Research: Conduct regular market research to understand trends and adjust your marketing strategies accordingly.
- Diversify Sales Channels: Explore multiple sales channels, such as local markets, restaurants, and online platforms, to reduce reliance on a single market.

Starting and operating a catfish farm presents a range of challenges, from water quality management to regulatory compliance. By being proactive and implementing effective strategies to address these challenges, you can enhance the success and sustainability of your farm. Continuous learning, careful planning, and adaptability are key to overcoming obstacles and achieving long-term success in catfish farming.

Encouragement for Your Catfish Farming Journey

Embarking on a catfish farming venture can be both exciting and challenging. As you take your first steps into this rewarding field, remember that success often comes from perseverance, careful planning, and a passion for what you do. Here's some encouragement to help you stay motivated and focused on your journey:

1. Embrace the Learning Curve

Starting a catfish farm is a learning experience, and it's natural to face challenges and make mistakes along the way. View these moments as opportunities for growth. Every problem you solve and every lesson you learn will bring you closer to mastering the art of catfish farming. Stay curious and open to new information, and don't hesitate to seek advice from experienced farmers and industry experts.

2. Celebrate Small Victories

Each milestone, no matter how small, is a step forward in your farming journey. Celebrate achievements like successfully stocking your first batch of fingerlings, reaching growth targets, or improving water quality. Recognizing these successes will boost your confidence and keep you motivated.

3. Stay Passionate and Committed

Your passion for catfish farming will drive you to overcome obstacles and find innovative solutions. Stay committed to your vision and remember why you started. Your dedication and enthusiasm will not only help you succeed but also inspire those around you.

4. Build a Support Network

Connect with other catfish farmers, join industry associations, and participate in online forums or local groups. A strong support network can provide valuable insights, encouragement, and practical advice. Sharing experiences with fellow farmers can also offer emotional support and reassurance during challenging times.

5. Focus on Sustainability and Innovation

Embrace sustainable practices and look for ways to innovate within your farm. By focusing on environmental stewardship and efficiency, you not only contribute to a healthier planet but also create a more resilient and profitable operation. Innovation can set you apart and open new opportunities in the market.

6. Learn from Setbacks

Setbacks are a natural part of any farming venture. Instead of being discouraged, use them as learning experiences to refine your practices and improve your farm's resilience. Analyse what went wrong, adjust your strategies, and come back stronger.

7. Seek Continuous Improvement

Strive for continuous improvement in all aspects of your farm, from water quality and feed management to marketing and sales. Stay informed about the latest industry trends and advancements, and be open to adopting new technologies and methods that can enhance your farm's performance.

8. Remember Your Impact

Catfish farming is not just about producing fish; it's also about making a positive impact on your community and contributing to the food supply chain. Your work supports local economies and provides a valuable source of protein for consumers. Take pride in the role you play in this important industry.

9. Maintain a Positive Attitude

Your attitude will greatly influence your success. Stay positive, even in the face of challenges. A positive mindset helps you stay focused on your goals, motivates you to keep going, and encourages you to find solutions rather than dwelling on problems.

10. Enjoy the Journey

Finally, enjoy the journey of building and growing your catfish farm. Take time to appreciate the progress you've made, the skills you've developed, and the experiences you've gained. The journey itself is as valuable as the destination.

Starting a catfish farm is a significant undertaking, but with dedication, resilience, and a passion for aquaculture, you can turn your vision into a thriving reality. Embrace each challenge as an opportunity to grow, stay motivated, and celebrate your successes along the way. Your hard work and perseverance will pave the way for a rewarding and successful farming venture.

www.ingramcontent.com/pod-product-compliance
Lightning Source LLC
Chambersburg PA
CBHW050258230526
45471CB00005B/1928